# Oilfield Trash

Number Twenty-two:
KENNETH E. MONTAGUE SERIES
IN OIL AND BUSINESS HISTORY
*Joseph A. Pratt, General Editor*

# Oilfield Trash

## Life and Labor in the Oil Patch

Bobby D. Weaver

TEXAS A&M UNIVERSITY PRESS
*College Station*

First edition

This paper meets the requirements of
ANSI/NISO, Z39.48–1992
(Permanence of Paper).
Binding materials have been
chosen for durability.

Library of Congress Cataloging-in-Publication Data

Weaver, Bobby D. (Bobby Dearl), 1936–
Oilfield trash : life and labor in the oil patch / Bobby D. Weaver. — 1st ed.
p. cm. — (Kenneth E. Montague series in oil and business history ; no. 22)
Includes bibliographical references and index.
ISBN-13: 978-1-60344-205-3 (cloth : alk. paper)
ISBN-10: 1-60344-205-7 (cloth : alk. paper) 1. Petroleum workers—Texas—Social life and customs—20th century. 2. Petroleum industry and trade—Texas—History—20th century. I. Title. II. Series: Kenneth E. Montague series in oil and business history ; no. 22.
HD8039.P42W43 2010
622'.182820976409041—dc22
2010002450

*To the memory of Robert S. "Bob" Weaver*
*The best damned tank builder in West Texas*

# Contents

# List of Illustrations

# Introduction

Between the first gigantic oil discovery at Spindletop in 1901 and the last bona fide oil boom in West Texas during the 1950s, the story of the Texas oilman achieved legendary status. During that time the flamboyant image of super wealthy wildcatters downing twenty-year-old bourbon over at the petroleum club became the general public perception of the oilman. Few gave a passing thought to those oilfield hands in greasy overalls who made the boomtowns boom. Those workers performed the hard, dirty, and often dangerous jobs of a specialized nature that created the oilfields of the Lone Star State. Very few of them ever saw the inside of a petroleum club.

This is the story of those blue-collar workers. Although this book focuses on activities in Texas it is indicative of the nature of oilfield workers everywhere. It explores their origins, the nature of their work, their lifestyle, and how all that changed over time. Technological innovation combined with influences imposed by the changing geographic location of oilfield development modified the nature of the work, although the underlying culture of oilfield labor remained little changed over time. The very product those oil workers produced caused the internal combustion engine to become the force that revolutionized transportation. It also spawned the development of specialized machinery that changed the manner in which they labored. Ironically, those same innovations ultimately caused the demise of the phenomenon of the oil boomtown and ended a number of traditional oilfield occupations.

The oil booms in which those workers labored were not all created equal. Some appeared on the marshy Gulf Coast and others in the dry sandy reaches of West Texas. Some developed in reasonably well-populated areas and others in almost deserted regions. Some were based around well-established communities and others at newly established towns. Some began during a time of minimal mechanical capabilities and others during an era of sophisticated technology. But all of them were populated by a young and rambunctious lot drawn from a rural background who, through their rough-and-ready ways, earned the sobriquet of oilfield trash from those outside the industry. It is difficult for solid, sober-minded citizens living a settled nine-to-five lifestyle to understand those who, for all practical purposes, have no permanent job and wander from place to place while earning a living under trying circumstances. At the same time, oilfield work attracted many from that same settled group who were dreaming of a more adventurous way of life.

The focus of this book is on the "upstream" contractors, or "boomers," who created the booms when they rushed in to develop the oilfields. They are the "oilfield trash" so vilified by locals and other outsiders who viewed the unruly nature of those mostly single, young oilfield workers with a jaundiced eye. This is not the story of the "downstream" company men, refinery and petrochemical workers, and professionals such as geologists and engineers who operated the oilfields in a more settled lifestyle after the booms ended.

The comings and goings of oilfield labor created a fluid social structure that was very egalitarian in nature. After all, with a little luck any one of them could become an oil baron. However, allowing for great overlap and blurring of lines, the generalization of professional management, oil company workers, and contractors/boomers fairly well defines the social structure of the oil fraternity as it developed over the first fifty years or so of the twentieth century.

Among the contractors/boomers there was a pulsing, push/pull effect that attracted workers to the oilfield. During a boom, which always demanded a great deal of labor, many locals were attracted to the work by the high wages. As the boom subsided, usually after one to three years, many of them returned to their former occupations, but some continued in oilfield work and moved on to the next boom. Over time that group formed the core of the boomers. The continuous need for labor was also kept high as boomers married, started families, and opted to settle down and work for the major oil companies who offered a more stable economic situation. Despite performing almost identical types of labor, this tended to set those company men apart as having a slightly higher social status than those still following the booms. Their lifestyle also served as a contrast to the colorful reputation of the so-called oilfield trash.

My family moved to the West Texas oilfields when I was twelve years old. My dad was a tank builder, and as soon as I was big enough to tote iron I went out in the crew. I spent more than twenty years working at that trade. Even before I went out on my first tank job I was fascinated by the stories told by the older oilfield hands. Theirs were adventurous yarns about oil booms at places with exotic names like Smackover, Hogtown, Wink, Borger, and Kilgore. They spoke mostly of good money, bad whiskey, and tough people who worked hard and lived fast during exciting times. That was heady stuff for an impressionable young person.

When I entered college, much later in life than most, I took those stories with me, and I added to them through my oral history work at the Southwest Collection at Texas Tech University. Ultimately, I discovered other treasure troves of serious oilfield related oral history collections at places like The Center for American History at the University of Texas; the archives at the Panhandle-Plains Historical Museum at Canyon, Texas; the archives at the Permian Basin

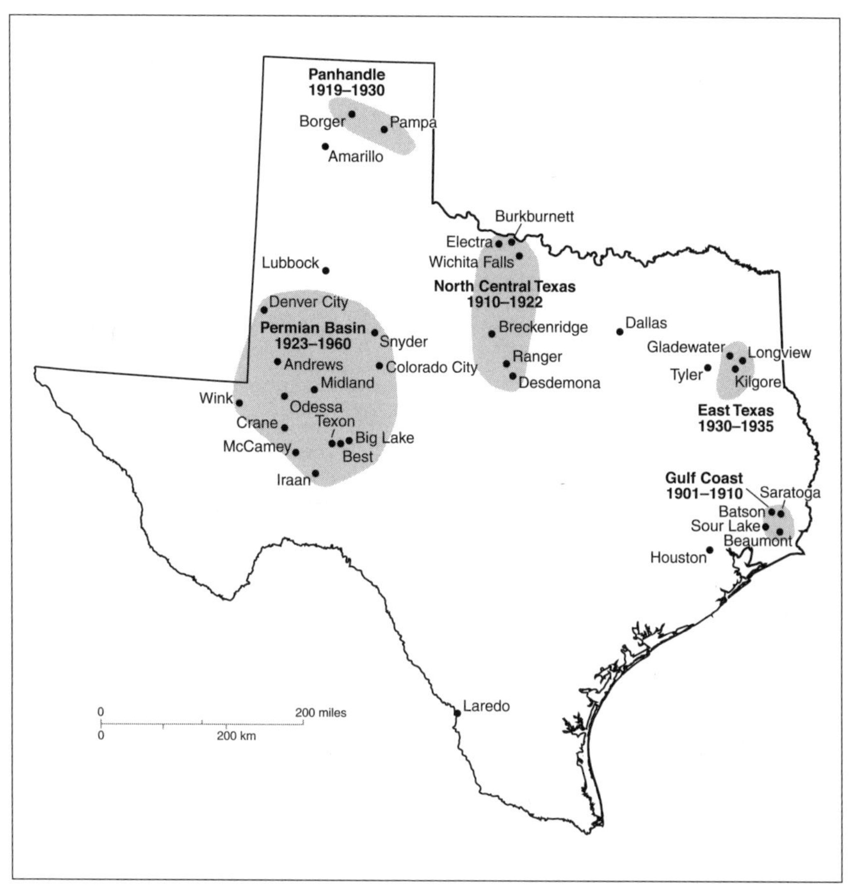


*The areas and time frames where the oilfield trash discussed in this book lived and labored.*

Petroleum Museum at Midland, Texas; and at the East Texas Research Center at Stephen F. Austin University at Nacogdoches, Texas. All told, I collected approximately 350 of those interviews that represent the nucleus of this book.

There are those who distrust the anecdotal nature of oral history to provide a solid basis for serious historical study. This is largely true in respect to pseudo-factual information such as "it was actually John Doe and not George Smith who shot Bill Jones on the morning of June 6 some fifty years ago." You might question the veracity of that sort of unsubstantiated information. However, when someone proudly explains how a complicated job was performed, what living conditions were like, or some other everyday lifestyle situation of a bygone era, you can be pretty sure that was the way it happened. By comparing large numbers of those narratives over time and through space, patterns of a lifestyle and how that lifestyle changed over time and in different locales emerge. As those patterns become apparent and are supported by contemporary newspaper accounts, industry publications, and serious academic studies, a fascinating lifestyle, enunciated by those who lived it, comes into focus. The evolution of that lifestyle says much about the nature of contemporary Texans.

In piecing together this oilfield odyssey I have largely ignored refinery and petrochemical workers as well as the employees of the major oil companies. Their more settled way of life serves as a contrast to that of the contractors or boomers who performed most of the work in establishing the oilfields of Texas. In that same vein I make no attempt to chronicle every boom in the state. Those areas explored represent different geographical regions in the general chronological order of their development. In this manner both technological and social change in the oilfield can be better understood as it developed over time and under different geographical influences. In general this book is an attempt to describe the work performed by various oilfield occupations, how those jobs changed or even disappeared over time, and the lifestyle surrounding those who performed that labor.

# Oilfield Trash

*The oil business in Texas got off to a running start on January 10, 1901, when the Lucas #1 came in at 100,000 barrels per day and blew wild for nine days. (Photo courtesy of Southwest Collection/Special Collections Library/Texas Tech University, C. C. Rister Collection)*

CHAPTER 1

# The Oil Boom, 1901–1905

The twentieth century arrived slightly later in Texas than in most places. The activity destined to dominate the state's social and economic life for the century was delayed for slightly over a year. Seldom does a single event characterize an entire century, but in Texas destiny intervened on January 10, 1901. That was the day the Lucas #1 blew in at Spindletop and propelled the Lone Star State from its legendary cotton and cattle past into a petroleum-dominated twentieth century.

Few native Texans had a working knowledge of the oil business at the turn of the twentieth century, nor did they have any idea of how to deal with the associated boomtown environment that activity brought. Their only experience with petroleum, prior to the 1901 Spindletop phenomenon, lay with a minor discovery near Corsicana in 1894. That happened when the city of Corsicana contracted to have three municipal water wells drilled. At a depth of 1,027 feet the drilling contractor encountered a major nuisance in the form of oil, which forced the driller to case the well in order to safely complete his contract to provide uncontaminated water to the city.

Immediate local interest resulted in the organization of the Corsicana Oil Development Company. That firm brought in a couple of experienced Pennsylvania oilmen named J. M. Guffey and John H. Galey to help drill and develop the wells. By 1897 a small oil boom was underway as activities attracted hundreds of oilmen from Ohio, Pennsylvania, and other oil-producing regions back east. Sleepy little Corsicana became the first Texas oil boomtown where boarding houses filled to capacity, all types of rental property became occupied, and local businesses experienced unprecedented growth.

The activity soon exceeded the local economy's ability to utilize petroleum production, and by late 1897 Corsicana oil was selling for as little as fifty cents per barrel while oil from the older eastern fields brought double that amount. At that juncture local investors persuaded yet another Pennsylvanian, J. S. Cullinan, to assist in the development of their little five mile by two mile oilfield. Cullinan persuaded all the local producers to sell their production to him and then proceeded to build the first permanent oil refinery in the state. By late 1898 Corsicana oil was back up to a dollar per barrel, and Cullinan was busy developing new ways to utilize the product. His ideas ranged from promoting it as a new locomotive fuel to coating the dirt streets of regional towns in order to keep the dust down.[1]

While modest success smiled upon Corsicana's efforts, other Texans were busy looking for similar rewards. The unlikely candidate for developing petroleum prospects on the Gulf Coast became Pattillo Higgins of Beaumont. Possessing only a third grade education and having lost an arm as a youngster, Higgins was regarded in his hometown as something of an eccentric. Through late night study over a lengthy period of time he developed a working knowledge of geology that gradually grew into the certainty that there was oil in the immediate vicinity of Beaumont. The particular focus of his obsession became a twelve- or fifteen-foot rise of ground located about four miles south of town known variously as Sour Spring Mound, Spindletop Hill, or just the "hill." It was not too difficult to imagine that petroleum existed beneath that hill when only a casual examination revealed that oil oozed to the surface in several places while gas bubbles occasionally burst from the surface of the stagnant ponds that abounded in the area.

As early as 1892 Higgins had persuaded some local businessmen to form the Gladys City Oil, Gas, and Manufacturing Company in order to investigate the oil seepages at Spindletop Hill that so intrigued him. Over the next four years the company made three attempts to drill oil wells at that location. Using traditional cable tool drilling rigs, their efforts were thwarted each time by an unconsolidated formation they called the heaving sand located a few hundred feet beneath the surface. Then in 1899 Higgins, by now only a minor partner in the company, but still the main force behind oil exploration in the area, persuaded A. F. Lucas to assume the lease on the oil property. Lucas, an experienced salt dome driller, had been operating successfully in Louisiana for several years. His first attempt at drilling on Spindletop ended at a depth of 575 feet when the money ran out, although they had a small showing of oil. With the project seemingly at an end, Lucas and Higgins approached the head of the Texas Geological Survey for advice on how best to proceed. The geologist introduced Lucas to J. M. Guffy and John H. Galey, who had experienced success in the Corsicana endeavor. Guffy and Galey agreed to finance the enterprise for a major percentage of the lease.[2]

At that point the Pennsylvania oil financiers hired the Hamill brothers of Corsicana to drill the proposed well to a contract depth of 1,200 feet. The crew that arrived on Spindletop Hill about the first of October 1900 consisted of Henry McLeod, rig builder; Al Hamill, contractor; Peck Byrd, fireman; and Curt G. Hamill, driller. They traveled to Beaumont by train where they hired an African American man as cook, loaded a wagon with provisions, and drove the short distance to Spindletop. Lucas furnished them with a small plank shack with no window screens close to the location as a place to bunk and eat. There, under primitive living conditions and enduring unending mosquito attacks, the work began.[3]

The drilling of the Lucas well serves as an excellent example of early Texas oilfield lifestyle and the nature of the men involved. The Hamill boys, who had arrived in Texas from Pennsylvania as infants, were considered native Texans.

They received their well-drilling training in both cable tool and the newly developed rotary method while working in the Corsicana field. Most of their oil well drilling knowledge was derived from crusty old West Virginia and Pennsylvania cable tool drillers brought in during the Corsicana boom. Beyond what they learned from the older men their greatest asset lay in the ability to overcome the technological problems peculiar to the oil business by applying ingenious practical solutions to essentially straightforward issues. The equipment they used, the best available at the time, had seldom been tested on well depths exceeding 1,500 feet.[4]

As Curt Hamill stated in later years:

> We was all farmers and we'd been used to crude machinery. And when we went into the oilfields, we found everything was crude there, and we started from the grass roots with crude machinery. Everything was hard to do. We had to use perseverance and we had to use regular old horse sense.[5]

The first problem they encountered on the Lucas project was how to build the derrick. As it turned out, Henry McLeod, the self-proclaimed rig builder, knew no more about building a drilling derrick than any other member of the crew. He was more of a hindrance than a help, which soon caused him to leave the job. The crew knew that the derrick was supposed to be sixty feet high, twenty feet square at the base, and five feet square at the top. Beyond that they did not have a clue. At that point they utilized the method used back in West Virginia where the drilling crews constructed their own derricks on site from local stands of timber. They solved the problem by getting fresh-cut lumber from a nearby sawmill, laying the green wood out flat on the ground in the configuration they needed, and sawing the various pieces to the proper length. Then they simply nailed the ready-cut pieces together until they had a finished derrick assembled. It took them about ten days to erect the derrick and set up the steam-driven rotary drilling equipment chosen for the job. Thus, in late 1900, drilling began utilizing the newly invented rotary method that had been in use only about five years.[6]

The Hamills spudded in around mid-October and quickly made 160 feet of hole. At that point they encountered the heaving sand that had been the nemesis of all previous cable tool attempts. After repeatedly losing the circulation of the clear water used as a drilling fluid necessary to wash the cutting out of the hole in rotary drilling they abandoned the regular drilling process and resorted to a pile driving technique. The crew rigged up a weighted device as a driver and pounded eight-inch casing into the soft sand a few feet at a time. Every few feet they stopped, rigged up their water pumps, and washed the sand out of the inside of the casing and then repeated the entire process again. Using this clumsy procedure the determined drilling contractors gained a few laborious feet each day. That exhausting, man-killing work continued for about three weeks, during which time they passed through 285 feet of the soft heaving sand. At 445 feet they struck "gumbo," as the local shale formation was known, and were able to resume normal drilling operations.

With the heaving sand obstacle out of the way the project proceeded fairly well until early December when they struck yet another coarse sand formation at a depth of between 640 and 700 feet. Once again the sand absorbed the water they were using as a drilling fluid and once again they lost circulation. After wrestling with the problem for several days they ran out of their supply of pine slabs used to fire the boiler, and at the same time the water well supplying drilling fluid went dry. At that point Al Hamill ordered drilling to cease while he went to secure more fuel and to repair their dulled drill bits. The crew, which by this time had dwindled to Al's brother, Curt, and Peck Byrd, was left with orders to clean the rig and prepare for the resumption of drilling upon the contractor's return.

During the three days Al was gone from the location Curt and Peck developed a process destined to have a profound influence on oil well drilling for all time to come. They were sick and tired of the long hours and extra work associated with losing circulation in the porous sand formations, and most of their conversations centered on solving that pesky problem. Both men had noticed that while they were drilling through the gumbo the clear water drilling fluid had gotten muddy, which seemed to help improve the circulation of the drilling fluid. After kicking the idea around for a while they decided that if they could make the regular drilling fluid into a thin mud it might seal off the porous sand around the edges of the hole and stop the lost circulation problem. With that thought in mind they approached the local farmer they had hired to keep the sand cleaned out of their slush pit to see if he could help them develop a muddy water supply. After cleaning the pit, the contractor, a local farmer named Reverend John C. Chaney, plowed the clay bottom of the pit to a depth of ten or twelve inches. Then, after running a foot or so of water into the pit, Chaney drove his herd of cattle into the water and walked them back and forth for about four hours until the liquid became thick with mud. By the morning of the third day of Al's absence the crew was adding water to the viscous mixture, which was being circulated by the drilling fluid pump to assure a proper consistency.

Late that afternoon when Al Hamill returned, he was, to say the least, skeptical of the benefits of his crew's new idea. The contractor was more inclined to put his faith in the modified drilling bit that he had instructed a blacksmith to fabricate during his absence. After much wrangling they agreed to try the newfangled "mud" in addition to Al's sharpened bit. When drilling resumed they lost almost half of their reserves of drilling mud with no fluid circulation evident before the new process began to work. As the mud began to seal off the sand and create a clay sheath on the edges of the hole, progress picked up significantly and with one day's work they drilled through the sand and back into the gumbo. The invention of drilling mud by the Hamill crew probably saved the Lucas well and assured the oil well drilling industry of one of its most useful tools from that time forward. By Christmas Eve they had gotten down to 920 or 930 feet, set six-inch casing after getting a showing of oil, and stopped the operation in order to return to Corsicana for Christmas. After three months of backbreaking labor the drilling crew was exhausted.[7]

The two Hamill brothers and Peck Byrd returned to Spindletop on New Year's Day, 1901, and resumed work on the well. This time, however, Curt brought his wife and four children with him. Their initial experience of living in a shack without screens, sleeping beneath double mosquito nets, and enduring less than good meals while working long and exhausting days had convinced all concerned that they needed better living conditions. Accordingly, Curt set up living quarters in a house about halfway between Beaumont and the well, and Mrs. Hamill cooked for the crew. That arrangement made the situation a little more bearable as they strove to complete drilling the well to the contracted depth of 1,200 feet.

Work on the well went smoothly for three or four days as an additional one hundred feet was completed. Then trouble developed again. At a depth of between 1,000 and 1,020 feet the bit stopped turning. It appeared to be wedged in a crevice of some sort. After worrying with the problem for two or three days and trying every trick they could devise, the crew could not overcome the problem. Finally, Al sent to Corsicana for a different type of bit to see if that would get them going again. The new bit arrived on the morning of January 10, and the crew immediately attached it and started lowering the drilling string into the hole. As the bit passed the 700-foot level, drilling mud suddenly began to boil up out of the well and the drill pipe began slowly rising of its own accord. Curt, who was in the derrick at the time, managed to scramble down to the drilling floor and shut down the boiler fires to prevent the danger of a well fire. He ran away from the rig as fast as possible to join the rest of the crew that had already beat a hasty retreat to a safe haven some distance from the developing phenomenon. As they watched in amazement, the rising drill pipe gained momentum until all 700 feet shot out of the top of the derrick, broke into segments, and fell back to earth in a tangled heap of twisted metal. None of them had seen anything like that before.

The destruction ceased in a few minutes. Then the crew cautiously approached the scene to view the carnage. The traveling blocks were wrecked, the crown block had been blown off the derrick top, pipe was scattered all over the landscape, and drilling mud covered everything to a depth of six inches. As they stood there surveying the scene they were scared out of their wits when a loud boom heralded the expulsion of a solid column of mud from the well. That was immediately followed by a quantity of blue gas, which subsided very quickly.

Then the crew, which by this time had made it to the drilling floor, heard a gurgling sound and oil began to flow out of the well and over the rotary table. The oil ebbed and flowed as the gas pressure gave it the appearance of breathing, and with each breath the flow grew higher and higher until finally in a mighty surge it shot above the derrick and sprayed across the countryside. The most famous oil well in the United States had just blown in.[8]

The well came in around eleven o'clock on the morning of January 10, 1901. By mid-afternoon the entire population of Beaumont, Texas, that was able to make it to the location was milling around gawking at the well that was still blowing wild. Al quickly notified Captain Lucas who just as quickly wired Mr. Galey in Pennsylvania that their well had been successfully completed at an unprecedented rate of flow. The news spread rapidly, and within a couple of days thousands thronged to the area. The well was estimated to be flowing at a rate in excess of 100,000 barrels per day in an era when production of one hundred barrels per day was considered good. It opened up a seemingly unlimited supply of oil and inaugurated the modern petroleum era to the world.

The Lucas well blew wild for nine days before the Hamill boys were able to get her capped. The area surrounding the well became flooded with hundreds of thousands of barrels of oil, and the well site became flooded with thousands of onlookers. The curious Beaumontites of that first day were a harbinger of the throngs that descended on the place from all over the country. Excursion trains were organized that brought untold numbers to the site, which quickly became a sort of magnificent tourist attraction. The visitors included curious tourists, serious oilmen, thousands looking for work, and speculators and promoters of every stripe.[9]

The leasing of the hill began immediately, and the unbridled development of the Spindletop find mushroomed beyond the original developers' wildest dreams. The two hundred acres that eventually produced oil was divided into three main sections, known as the Hogg-Swayne, the Yellow Pine, and the Keith Ward. Those rough divisions were each primarily controlled by a small group of people who did the main developing within its confines. For example, the Hogg-Swayne tract was developed by ex-Texas governor James Hogg and his partner, James W. Swayne. Those developers sold individual leases within their tracts as small as one-sixteenth of an acre, and many of those tiny allotments changed hands for as much as ten thousand dollars. By the end of 1901 as many as 585 oil and leasing companies were operating in the field, and the proliferation of the extremely small leases created a situation so crowded that the legs of drilling derricks actually touched one another at ground level. So much more money was invested in this new Eldorado than was actually recovered that one writer referred to it as "swindletop" rather than Spindletop. The town of Beaumont, a farming and sawmill center of approximately 9,500 inhabitants, exploded in size to an estimated 20,000,

*During the first months of 1901 excursion trains brought thousands to Beaumont to experience the carnival-like atmosphere surrounding Texas's first big oil boom. (Photo courtesy of Southwest Collection/Special Collections Library/Texas Tech University, C. C. Rister Collection)*

30,000, or even 50,000 souls by the end of the year. Those who flocked to the boom represented oilmen, businessmen of every description, promoters from all over the world, hordes of unskilled workers, and a black hovering cloud of the lawless, all attracted by the siren song of unbelievable wealth available for the taking.[10]

Sam Webb and Bud Coyle were typical of the working hands drawn to Spindletop during those first exciting months of 1901. They had both grown up in the Dallas area and were acquainted with Walter Sharp, whose Corsicana-born company was emerging as one of the major drilling contracting firms in the new field. Arriving a couple of months apart in the late spring and early summer, they both went to work for the Sharps in the Yellow Pine District. By that time men were pouring into the area in such numbers that living quarters were almost impossible to find. The two Dallasites were particularly fortunate to be working for the Sharps, who provided a camp where their employees stayed. The camp, which housed the forty or more men necessary to operate the six Sharp rigs running in that section, was considered the best available. It consisted of several wooden-floored tents raised twelve inches above the soggy ground. Each tent slept eight men. Adjacent to the tents was a wooden kitchen and a wooden dining hall. The men were fed good meals, and their quarters were cleaned daily by people hired especially for that purpose.[11]

Nineteen-year-old Frank Redman arrived about the same time. He entered Beaumont at 4:00 A.M. on June 10, 1901, as one of several travelers hiding in a boxcar. Redman spent his last twenty-five cents on what would normally have been a ten-cent can of sardines for his breakfast and walked the six miles to Spindletop. He got there at noon, and by one o'clock he had hired out to the Drummer's Oil Company as a roughneck. That firm, much like the Sharps' firm, provided tents for their hands' sleeping quarters, but meals had to be taken at one of the many boardinghouses that dotted the landscape.[12]

Not all the unskilled workers who descended upon the area were fortunate enough to find stable employment with oil companies and drilling firms right away. Some, like Claude Deer, who arrived in late August of 1901, had trouble finding steady work or even a place to sleep. For over a month he slept in a livery stable office that belonged to a friend from his hometown, and he ate wherever and whenever he could get a handout. Deer was indeed fortunate to have a protected inside place to sleep, for it was an exceptionally wet summer characterized by heavy rains that turned the streets into quagmires. The situation was so uncomfortable that men paid $1.50 to $2.00 per night to sleep on covered porches. Deer finally got steady work on the drilling rigs, but he led a precarious existence during his first few months in the oil business.[13]

Work on the drilling rigs was a well-paid job for the average working man at the turn of the century. The drilling business worked around the clock on two twelve-hour shifts that ran from twelve noon until twelve midnight. Workers were paid on the basis of a daily wage instead of an hourly rate. For example, drillers drew in the $5.00 per day range and roughnecks got $2.50 to $3.00 per day. General labor in the field ranged from $1.50 to $2.50 per day depending upon the type of work performed. Oilfield work was considered extremely dangerous, and all hands signed what was generally called a "death warrant." It was a waiver that simply stated that the person signing it was a white male over twenty-one years of age who assumed full responsibility for any accident that might befall him during the course of his employment.[14]

Housing, feeding, and entertaining all those who made the Spindletop boom presented a major problem for limited local resources. In general, Beaumont and the area of the hill assumed separate roles. The section of Spindletop not covered by drilling rigs became a sprawling mass of rooming houses, boardinghouses, saloons, grocery stores, and a myriad of other establishments designed to service the workingmen forced to live near their jobs. There were three small, designated towns that existed there during the boom, but they disappeared shortly after the main activity ended. They were Gladys City, Spindletop, and South Africa or Little Africa, where the African American community was confined. Beaumont, with already existing facilities, developed as the supply headquarters, administrative center, and financial center, and it also provided a convenient red light district. Thus, in the chaotic hysteria of Texas' first great oil boom, the area around Spindletop emerged as a temporary tent and shack community while Beaumont, with its more stable background and access to the railroad, assumed an orientation toward permanency and solid growth. In general this was a pattern that would persist throughout the next thirty or forty years as one oil boom followed another in the state of Texas.[15]

During the years of 1901 through 1904, when the Spindletop boom was at its height, an estimated 20,000 men were continuously at work on the hill. To serve their many needs a variety of commercial establishments appeared. Perhaps the most common were the rooming houses, which could sleep 100 to 125 men each. Those flimsily built structures were little more than partitioned barns with cots for sleeping. Many of them even rented their quarters on a twelve-hour shift basis in order to double their income. Most of the establishments charged in the neighborhood of $1.00 per shift for their beds. The rooming houses, along with company camps, provided most of the living quarters for single men in the field. The few families who went there either lived in floored tents or in hastily constructed shacks scattered along the flanks of the hill. The few structures that existed prior to the coming of the boom, such as the house Curt Hamill had, were soon taken up by the more affluent of the newcomers.[16]

Probably the second most common structures built on the hill were the boardinghouses. Those buildings were many times little more than sheds or half-walled tents with window screens between the roofline and the half wall. Meals were served on long plank tables with the customers seated on benches. The food, although not the best, could be provided for about seventy-five cents per day for three meals. Dining at one of those emporiums, which seated up to two hundred men at a time, was not for the fastidious. Flies abounded and hordes of mosquitoes bombarded the hungry oilfield hands as they ate in the noisy company of their fellow workers clad in oil-stained work clothes.[17]

One of the more enterprising businessmen in the area opened a soup kitchen that soon became one of the more popular eating establishments. This fellow heated water over an open fire in a series of vats four feet wide by six feet long by two feet deep. Then he dumped hundred-pound sacks of ground beans into the vats, which produced a soup that sold for fifteen cents per bowl. Seating two hundred to three hundred patrons at a time, his establishment regularly served 5,000 to 10,000 bowls of soup each day.[18]

Once the men at Spindletop were housed and fed, the next most important thing was entertainment; however, for many this aspect of their lives took precedence over food and shelter. Between 1901 and 1903 the number of saloons in Beaumont grew from twenty-five to eighty-one and they were scattered throughout the city. At Spindletop, entertainment usually took the form of visiting saloons. The twenty or thirty drinking establishments on the hill did not by any

*Tents and other temporary shelters served as places to feed the mass of people descending upon the Spindletop boom. (Photo courtesy of Southwest Collection/Special Collections Library/Texas Tech University, C. C. Rister Collection)*

means represent a stylish operation. They were commonly only tents or simple frame structures that sold beer by the bottle and whiskey by the quart. The whiskey, more often than not homemade corn liquor, sold for a dollar per bottle and was welcomed by all hands. Sometimes the saloons offered gambling opportunities, but very little prostitution existed on the hill. The drinking establishments tended to become the focus of most of the fistfights that took place in the area. It was not uncommon for men to be thrown through the screen wire walls of the saloons during these brawls. One eyewitness described seeing ten men fighting outside one of the joints on one occasion. Of the five fistfights in progress none had started for the same reason. One was over a girl, another was simply to see who was the best man, yet another was over a card game, and the remaining two appeared to be ordinary recreational activities. Those working at Spindletop were a young and a tough crowd who were definitely feeling their oats.[19]

Sooner or later the search for entertainment drew the oilfield hands to the red light district in Beaumont. That area, known as "the reservation," was described by Will Armstrong, who worked there as the night shift policeman from August 1902 until July of 1903. He remembered it as being three-quarters of a mile in circumference within whose confines eighteen saloons operated and approximately four hundred prostitutes plied their trade.[20]

The reservation saloons sported colorful names like Coney Island, the Klondike, Two Brothers, Little Casino, and the Blue Goose. One particular establishment attracted a clientele by having a live rattlesnake prominently displayed in a cage. When business slowed down the owner sent word around town that he intended to feed the snake. Men flocked to the saloon to see a live bird put into the cage. While everyone imbibed freely, lively betting developed on the length of the bird's survival. After the bird became a meal and all the bets were settled, the saloon owner enjoyed another hour or so of prosperity as the excited patrons discussed the bizarre event they had just witnessed.[21]

Large numbers of girls were employed by the saloons. They received a 25 percent commission on all the liquor they could sell. With beer at a dollar a bottle and champagne at five dollars per bottle a persuasive lady could do very well indeed. One memorable night Hazel Hoke, one of the more successful damsels of the evening, sold $1,300 worth of champagne to one man. He bought the bubbly by the case, but the proprietors only opened a bottle or two out of each case before trotting out a new one. By daylight the poor guy was so broke that he had to borrow money from a friend for train fare back to his home in Oklahoma. Another good source of income for the saloon girls lay in selling keys to their rooms. Depending upon how drunk the customer was, they could get from one to five dollars for each key. Some of those girls are credited with having a stock of forty or fifty keys to supply the demand they created. Although most of the saloon girls did work as part-time prostitutes, the reservation boasted significant numbers of full-fledged brothels. Beaumont had an ordinance prohibiting women from being on the streets after eight in the evening in the reservation area of town. Thus, the prostitutes were barred from traditional streetwalker activities in the red light district. The larger bawdy houses contained fifteen to twenty rooms and their madams were well known to the oil fraternity. Perhaps the most famous of these entrepreneurs were Hazel and Effie Hoke, who both sported thousands of dollars worth of jewelry that they wore everywhere they went. Almost as well known was Ruby Belle Pearson, who also operated a large house with the able help of her daughter. Most of the reservation madams hailed from New Orleans, although there was a sprinkling of sporting girls from across the United States. In addition to the larger and well known houses of prostitution there were thirty or forty crib houses in the reservation's confines. The cribs were long narrow structures with a series of doors opening along a board sidewalk that opened into small rooms just large enough to hold a bed. Then there were scores of the ever present "off path" one- and two-girl operations that catered to the oil boomers. Additionally, there were even a couple of African American houses, one run by Gold Toothed Sadie and the other by Big Annie, that harbored eight or ten girls each. The ladies in the larger more opulent bawdy houses commanded $2.50 for their services, while the African American and the French girls in the off path establishments received only one dollar for similar labor.

The wide open atmosphere of prostitution, gambling, and related criminal activities in the reservation area occurred under the watchful eyes of the local law enforcement officials who were determined to keep the boomtown vice

*The Hogg-Swayne tract at Spindletop was typical of well spacing during that boom when derricks stood leg to leg, as shown here in 1903. Drilling was evidently still proceeding, as indicated by the derrick being built in the foreground directly behind the cypress wood tanks. (Photo courtesy of Southwest Collection/Special Collections Library/Texas Tech University, C. C. Rister Collection)*

concentrated in that area. The Beaumont police force utilized a large wagon that they called the Black Moriah to transport wrongdoers from the reservation to the town jail. It sported four-foot-high sideboards and a rear door. Each night the most unruly drunks from the reservation were unceremoniously piled into the Black Moriah and hauled to their place of incarceration, usually to be fined and sent on their way the next morning. Except for the occasional shooting or accidental death, friendly fistfights excluded, violence on a massive scale did not exist. The primary law enforcement problem consisted of keeping boisterous drunks from spilling over into the more sedate portions of town.[22]

Spindletop occupied everybody's attention during those first few months of reckless activity. The population exploded as hundreds of experienced oil well drilling contractors brought their crews from the West Virginia/Pennsylvania fields to combine with the thousands of unskilled workers from the small towns, farms, and ranches of Texas and Louisiana. Soon those with prior knowledge of the business began to expand their search for the elusive black gold. Geologically, Spindletop represented a salt dome. It was simply a buried salt plug forced upward by geological forces that also allowed oil and its associated gases to accumulate beneath the higher portions of the umbrella-like top of the dome. Spindletop Hill represented the surface indication of a salt dome. Thus, any hill on the coastal plain of the Gulf of Mexico within a reasonable distance of Beaumont became suspect of harboring more oil riches.

As the oilmen fanned out across the region, Sour Lake in Hardin County, only twenty miles or so from Beaumont, became the site of the second oil boom on the Gulf Coast. The name Sour Lake was derived from the abundance of oil and gas seeps along the edge of a brackish lake. The mud from those seeps was reputed to have medicinal value. Accordingly, a small frame hotel had been built on the site years before and it did an excellent business catering to those who came to "take the cure" for their arthritis, rheumatism, and other assorted ailments. The best-known practitioner of healing with the sour mud in the area was a local black man known as Dr. Mud. Within a year of the Spindletop discovery oil scouts were trailing the mysterious Dr. Mud through the swamps and bayous around Sour Lake on a twenty-four-hour basis to discover just where the most prolific seeps lay.[23]

It did not take a genius to deduce that with oil seeping to the surface all over the place it was likely that a successful oil well could be drilled at the location. By early 1902 several small producers had been completed in the area. Walter Sharp became something of a legend for the part he played in starting the Sour Lake boom. Late in 1902 he drilled two fairly successful wells for the Texas Company in the immediate vicinity of Sour Lake, but he was able to keep the news from leaking out. In January of 1903, while drilling a third well, it became evident that it would become a gusher. Walter called the oil company people to the site, and around midnight, in the midst of a raging rainstorm, he brought the well in. It was a gusher in the tens of thousands of barrels range. They shut the well in, allowed the rain to wash the evidence away, and proceeded to negotiate an excellent marketing contract on the production from it as well as the rest of the wells that they had under lease. By this method the Texas Company managed to reap a tremendous profit when subsequent overproduction drove the price of oil down in the area.[24]

That massive discovery that touched off the Sour Lake boom early in 1903 brought hordes of the Spindletop workers to the scene. This time things differed significantly from the Beaumont situation. The production lay in a remote and swampy region accessible only by wooden corduroy wagon roads. There were no more than a half-dozen permanent structures in the vicinity. Nearly all the housing became tents built on or near the work site, which, once again, were the responsibility of the drilling companies like Sharp. Naturally, saloons opened as the first businesses on the scene, but they were short-lived because the boom fizzled in less than a year.

The nature of the Sour Lake field and the way it was drilled doomed its long-term prospects. As at Spindletop, drilling proceeded at a frantic pace. Once again rigs stood derrick leg to derrick leg. However, this time the salt dome formation covered a smaller area. Perhaps the most unusual element in the field was the Shoestring District where wells were concentrated in a narrow strip only forty or fifty feet wide and three-quarters of a mile long. The 150 wells in the Shoestring

District became known as Texaco row because all the wells belonged to the Texas Company. Massive overproduction dissipated the gas pressure necessary for the wells to flow so that within a year seventy-five of the 150 wells drilled in the Shoestring had to be abandoned.[25]

The poison gas problem that had also plagued oilfield workers at Spindletop developed into a major threat at Sour Lake. The hydrogen sulfide gas prevalent in the Gulf Coast wells proved deadly to those breathing it. In general, gas was so plentiful at Sour Lake that some producers actually drove short pieces of pipe into the ground to tap a small flow of the vapor. The vapor was set afire to serve as torches so that workers could see to walk to their jobs after sunset. From the very first activities in the area these fatal vapors caused several deaths and numerous near fatalities. Evidently, the high nighttime humidity in that region, combined with the large volume of gas, caused heavy concentrations at ground level, particularly in the low places.[26]

The Shoestring District was infamous for its problems with this phenomenon. Because the rigs there were so close together and large numbers of them were drilling at the same time, there was a constant danger from gas emissions. The situation was further complicated by the tangle of pipe, rig timbers, and other pieces of equipment that made it difficult for workers to evacuate in a timely manner if a well started venting gas. All rigs drilling in the Shoestring District, and most others in the immediate area, kept a lookout posted to warn the crew in case any neighboring well began to make gas. The situation became so difficult that Sam Webb, who worked as a driller in the Shoestring, related how he walked to his rig on a boardwalk across marshy ground, relieved the driller on duty, and worked as long as he could stand it in the gassy atmosphere. At that point he would relinquish his duties to a relief driller who would work as long as he could deal with it. Then the relief driller would turn the job over to yet another relief driller. Despite the efforts of the drilling contractors to maintain twelve-hour shifts, most of the crews around Sour Lake worked an irregular rotation much like that related by Mr. Webb. On one occasion, Early Deane, another Sour Lake driller, observed a man collapse from gas inhalation in the Shoestring. One of the hands volunteered to go in and bring the unconscious man out. The crew tied a rope around the volunteer's waist, whereupon he walked into the gassy area and brought the injured worker to safety. The injured party survived the incident after the volunteer applied artificial respiration. Those two accounts and others similar in nature typify the dangers associated with what was considered normal working conditions at Sour Lake.[27]

As the boom activities at Sour Lake began to wind down late in 1903 another spectacular find materialized at Batson, less than a score of miles away in an even more remote and marshy section of the Gulf Coast. The Batson boom represents the most lawless of the early Gulf Coast boomtowns. When the first big producer came in, there was no Batson. There was only a post office in a little country store at Batson's Prairie and no other dwellings other than the occasional isolated farm or ranch house. When word spread of the discovery late in 1903 a leasing frenzy developed and oilfield workers from Sour Lake and Beaumont swarmed into the area. Almost overnight a sea of tents appeared. William Cotton related that he was a few miles away in Saratoga on the day the well blew in. By the time he arrived at Batson's Prairie around noon the following day there were already seventy-five tents set up near the little store. Within weeks a two-block-long line of small frame houses appeared at the location and Batson was a reality. Housing was so scarce that witnesses reported seeing as many as 150 men sleeping in a giant circle around a large gas flare. With no other shelter available, they slept on the grass in the open and depended on the burning gas to keep them warm.[28]

Because no town existed prior to the boom, the county attorney, one Ralph Durham, assumed law enforcement responsibilities. Having had some experience at Sour Lake, Mr. Durham knew exactly what he wanted to do. He appointed a justice of the peace and hired several constables. Their primary duties seemed to consist of collecting fines every Monday from the gamblers and prostitutes who plied their trade in the town with impunity. In order to make this system work more efficiently, he devised what was locally called "the roundup." Every Monday morning the constables forced the saloon owners and madams to appear in court, where the saloon owners paid $25 for each gambling device in their respective establishments and every madam paid $12.50 for every girl working in her house. With thirty-three saloons operating in town, each harboring six to eight gambling devices, and an estimated two hundred girls working regularly, Mr. Durham and his cohorts garnered a considerable income.

The saloons were all of similar construction. They were simple frame buildings that contained a bar, a dance hall, a gambling parlor, and a stage room where vaudeville-type performances were presented. All those establishments remained open and sold liquor twenty-four hours a day despite Hardin County's statutes prohibiting the sale of alcoholic beverages. Business really boomed at night. The evening began early with vaudeville performances that usually ended around eight o'clock. Afterward, dancing commenced in the attached dance hall where the men paid fifty cents per set to dance with the girl of their choice. The girls received thirty cents of this fee plus a percentage of all drinks sold to their dancing partners. During the entire evening a gambling parlor, which included dice, cards, roulette, and other assorted gambling opportunities, operated at full throttle. All the while alcohol flowed freely. The saloons enjoyed a booming business, but attending them at night proved so dangerous for the working hands that they feared to venture into town alone. When the boys from a particular camp went to town they always went in groups of two to a dozen. By that time they had learned that it was the safest thing to do.

Because of the swampy nature of the Batson area it became almost impossible for the workers to travel the mile or so from town to the field. Accordingly, the oil companies operating during the Batson boom built a board walkway across the low-lying ground between the town and the field so their employees could get to work safely. The improvement was barely finished when an enterprising saloon owner built his business astride the walkway. By all accounts he did a thriving business as all shifts going on duty as well as those leaving work had to go through his establishment.[29]

Inevitably, the lawless situation at Batson attracted so much attention that state law enforcement officials were forced to take a hand. Early in 1904 the Texas Rangers were sent there to bring peace to the community. Their law enforcement was swift and sure. Ralph Durham and his appointees were quickly removed and assorted gamblers, prostitutes, and other lawbreakers were arrested. At first there was no jail, Durham and his crew had no need for one, so the felons, male and female, were chained to a convenient tree until they could be sent to the county jail. Soon, however, a jail was constructed of two-by-four lumber laid wide side down for strength, and the town had a place to incarcerate the lawless element. By 1905 the boom had subsided and the colorful types drifted on to the scene of the next boom.[30]

The first five or six years of the twentieth century were indeed exciting ones on the Gulf Coast of Texas. Spindletop ushered in an era of unbridled oil production that brought a completely new economic factor to the Texas scene. In 1903 the Sour Lake discovery, along with another boom at nearby Saratoga, expanded the activity. By 1904 Batson added to a mix that was soon followed by important discoveries at Humble, near Houston, which kept the excitement alive. These activities set the tone for all the major oil discoveries in Texas for at least the next thirty years in the way oil booms developed, in who participated, and in the lifestyle of those involved. Thus, during the first few years of the twentieth century oil production along the Gulf Coast created a craze that, as it developed and expanded outward, propelled Texas into the realm of legend.

The story of Texas oil is one of legendary multifaceted elements, but those elements ultimately come down to the tales of those who created that larger-than-life perception of the Texas oilman. Much has been said and written, and rightly so, of the entrepreneurs who created this unimaginable wealth, but the thousands who actually did the work, survived the exciting times of oil booms, and lived to relate their stories give the legend its life, color, and sense of reality. The hands in greasy overalls populated the boomtowns, did the work, developed the early technology, and made it happen.

CHAPTER 2

# The Drillers, 1901–1910

Perhaps the most glamorous work in the oil business is well drilling. The very idea of oil pouring out of a hole drilled into the earth ignites the flame of adventure in most young men's imagination. The added dimension of gaining great wealth heightens the romantic aura surrounding the process. However, to those workers caught up in the maelstrom of excitement surrounding oil booms, reality never quite develops into the romantic adventure it seems from afar. Nevertheless, the fast-paced life and good wages draw workers like the siren song described in the poetry of the ancient Greeks.

There are two basic ways to drill oil wells. Cable tool drilling, a percussion system whereby the bit pounds a hole in the earth, was the traditional method used to drill wells at the beginning of the twentieth century. By that time cable tool drilling had attained its maximum degree of technological innovation. Rotary drilling, the newcomer, first made its Texas appearance in 1895 in the oilfields at Corsicana and over the next thirty years was destined to become the primary drilling method used in the oil business. The rotary method of boring a hole in the earth with an augerlike motion worked well in the unconsolidated formations of the early Texas oilfields. That technology became the primary system used in the Gulf Coast oil booms.[1]

Rotary drilling, like all oilfield labor, involves strenuous, dirty, and often dangerous activity. Although the discovery well at Spindletop utilized a three man crew, the process soon settled down to five hands for operating a rotary drilling rig. Of the five men who made up a rotary rig drilling crew at the turn of the century, four were collectively called roughnecks, although within the group they were divided into a derrick man, two floor hands, and a fireman. The fifth was the driller, and he was the only skilled man absolutely necessary to the operation. The ability to complete a well a thousand feet or more beneath the earth set the driller apart as someone special. Drillers took their work seriously, assumed a superior demeanor, and brooked no interference in their domain.[2]

The few experienced rotary men at Spindletop in those first days had learned their trade at Corsicana. H. P. Nichols, who arrived on the scene in March of 1902, claimed that there were only five experienced rotary drillers at Spindletop. He allowed that they were the two Hamill brothers, the two Sharp brothers, and Jess Lincoln, all Corsicana products. Nichols went to Beaumont to turn his cable tool drilling experience to good advantage. Fortunately for him the Cincinnati insurance man who hired him knew little about the oil business and less about drilling rigs. He led Nichols to a pile of rotary drilling equipment lying on the ground in the Hogg-Swayne division and told him to assemble it and drill a well on a particular one-sixteenth of an acre lease. The newly hired driller, not having the foggiest notion of how the rig worked, began to hang around similar rigs that were running in the vicinity in order to learn the business. He eventually determined that the drilling fluid pumps, which he had never seen before, were simply converted steamboat pumps. With that bit of insight he set out to hire a crew. As he explained it:

> That night I made the rounds of the joints, and at one of these I bumped into a man who alleged that he knew all about pumps, that at one time he had been an engineer on the *Belle of Austin,* a pleasure boat operating on Lake McDonald, an artificial lake west of Austin, Texas. Augmented by this vast technical experience, he and I rounded up some six or eight additional men, two of whom had fired a boiler in an East Texas sawmill. Truly a motley crew: one ordinary cable tool driller, a steamboat engineer, and a bunch of fellows from the piney woods of Texas, who had been lured from their homes by a promise of three dollars per day for twelve hours of hard work. But three dollars per day in 1902 was a lot of money.

It took Nichols and his "motley crew" six months to drill that first well. He stayed with the work and after a few more jobs he was completing wells in about one week's time. Nichols went on to comment that "the qualifications of an early day driller was determined by the amount of liquor he could waddle around with, and the number of fistfights he could win."[3]

The shortage of experienced drillers at Spindletop created numerous situations similar to that of H. P. Nichols. Those men who came into the area with any kind of drilling experience, and many did from the Pennsylvania and West Virginia fields, were immediately hired to run the rotary rigs. However, most of the traditional cable tool men refused to have anything to do with the rotaries. Consequently, as soon as time allowed promising young workers from local areas were trained to assume drilling positions. Sam Webb is a good example. He began roughnecking for the Sharps in 1901. After working for a year or so he went to Sour Lake where within weeks of his arrival he was promoted to driller for the company. In Sam's situation, as with most drillers, the contractor simply told them to drill to the oil sand, set pipe, and complete the well. Most times they would only see the boss once or twice more before completing the well. Leaving those drillers to their own devices instilled a great sense of pride and independence in them as they experienced success at "bringing in wells." Before long that sense of pride was transformed into such an intense independent attitude that it became one of the overriding characteristics of early day oil well drillers.[4]

As young men left the farm and flocked to the oilfields a cadre of experienced rotary oil well drillers began to develop. When they first arrived in the "oil patch" the older workers called them "boll weevils" in reference to their agricultural origin, but as the new men began to "make a hand" they were accepted into the oil fraternity. Using that same frame of reference the hands began to refer to their workplace as the "oil patch," just as they would speak of a cotton patch or a corn patch back on the farm. Those references have stuck in the oilfield, although by the 1950s the term *boll weevil* became interchangeable with *worm* in reference to new hands.[5]

The temptation to move on to the next boom was almost irresistible to those young men. Bud Coyle is a typical example from those days. He moved around a lot. Coyle started with the Sharps at Spindletop in 1901 and moved with them to Sour Lake in 1903. The following year he drilled at Batson for a few months before moving on to Saratoga where he drilled a couple of wells. By late 1904 he was drilling at Humble near Houston where he stayed for more than a year before returning to Sour Lake. For the next year or so he bounced back and forth between Sour Lake and Humble, depending upon which drilling contractor needed his services. After that he did a stint in Louisiana and Alabama on some wildcat wells before returning to the Texas Gulf Coast. He stayed in that area in a variety of locations until 1910 or 1911 when he heard of a boom out in California. While preparing to go west an attack of fever laid him low. As he was recovering from the illness a friend talked him into going to North Texas to the new discovery at Electra, just west of Wichita Falls. He worked there, at Burkburnett, and drilled some unsuccessful holes in Young County before returning to Houston in 1912. Once there Coyle worked almost every field in the Houston-Beaumont area until 1920 when he formed his own drilling company. Thus, over a period of almost twenty years the man plied his trade in as many places while experiencing the exhilaration of taking part in several booms scattered across the country.[6]

The nature of the work the early drillers performed was very imprecise. Their crude drilling equipment assured that only a highly resourceful individual could operate the rigs with any degree of consistent success. The numerous problems associated with what was happening deep beneath the earth called for a special feeling for mechanical devices and what they were doing. Not all who attempted drilling oil wells had that gift, but those who succeeded were often regarded with awe by their fellow workers for the phenomenal intuition they exhibited and their cool demeanor under extreme pressure.[7]

The Hamill boys assumed a legendary status from the beginning. Drilling the discovery well at Spindletop while developing drilling mud in the process was accomplishment enough, but the feat that assured them immortality among the hands lay in their capping the Lucas well. When the well blew in and spewed oil all over the landscape no oilman had seen anything like it. Everyone was in a quandary as to how to stop the gusher. Offers to do the job poured in by the score from across the nation in prices ranging from $10,000 to $100,000. Finally, Mr. Galey gave the Hamills the opportunity to cap the well in view of the fact that they had drilled the thing in the first place. They devised a series of valves that in later years became known as a "Christmas tree" to be maneuvered over the wild well that, when put into place, could be closed off to stop the flow from the wellhead. However, before the apparatus could be screwed into place the thread protector on the casing protruding from the well needed to be cut off and new threads had to be cut into the exposed casing. The crew decided that such a dangerous job should be done by Al Hamill, the only single man in the crew. Clad in a rain slicker, rain hat, and wearing a tightly fitting pair of goggles, Al worked patiently all one afternoon just inches from the raging flow with oil raining down on him and gas hanging all about. Amid all that confusion he cut the thread protector off with a hacksaw and carefully reconditioned the exposed and damaged threads with a hand file. After Hamill was satisfied with the job the crew matter-of-factly eased their capping device, an object and procedure never before tried but destined to become a standard item in the oilfields, into place and closed the valves to stop the wild flow. Curt Hamill, who closed the valve to shut in the well, was overcome with gas and thus holds the distinction of being the first person to be gassed in the Texas oilfields.[8]

As it turned out gas proved to be a major hazard to drilling crews in the Gulf Coast region. There were numerous incidents and some deaths reported, especially in the Sour Lake area. One such event occurred at Sour Lake when a young man was left to guard an idle drilling rig overnight. His brother walked out to visit him during the course of the night and the next morning both were discovered dead from gas inhalation. Most gas victims survived when fellow workers dragged them to safety and applied a primitive form of first aid that consisted primarily of beating them on the chest to get them breathing again. However, all gas victims suffered from temporary blindness and some retained permanent vision problems from the experience. The hands developed a variety of remedies to treat the eye problems associated with gassing. The treatments usually consisted of covering the eyes with a poultice of some sort, such as scraped Irish potatoes or raw meat, although many preferred soaking the eyes with a strong saline solution. Blindness resulting from gassing lasted for varying lengths of time, ranging from a couple of days to a week or two and usually had no permanent effect.[9]

Along with the gas danger was the ever-present specter of fire. Given the technology of the day and the frantic rush to drill as many wells as possible in the shortest possible time using wooden derricks soaked in oil with gas spewing from numerous wells, fires became a constant threat. The first big blaze developed about a month after the completion of the discovery well. The hundreds of thousands of barrels of oil that escaped from the Lucas gusher had collected in

*This drilling crew is practicing an artificial respiration technique, ca. 1915, which indicates how dangerous the poison gas situation was to oilfield workers and how seriously they took the threat. (Photo courtesy Southwest Collection/Special Collections Library/Texas Tech University, C. C. Rister Collection)*

the flatlands nearby and soon afterward a particularly heavy rainstorm transformed the area into a lake of oil bordering on the railroad right-of-way. Curt Hamill witnessed the fire and described the event.

> It caught fire along in the middle of the day, we'll say, from this locomotive, and it burned until about four o'clock in the evening and got clear across that end of the lake. It was probably three quarters of a mile to the other end of the lake. So they thought it best to get rid of the oil; so they set fire to the other end of the lake so it would burn up quicker. When these two fires met, it created a great explosion. When the fire would meet it would throw hundreds of barrels of oil up in the air, and it'd explode in the air and plumb down to the ground. And just jarred the whole country just terrible. Well that would throw the oil back maybe a hundred yards, and it'd rush right back up and explode again. So it just played back and forth there for three or four hours thataway and people all over the country was scared to death.[10]

As the Spindletop boom and other operations developed over time there were many more fires. The drilling rigs became so closely spaced on Spindletop Hill that workers could walk from rig floor to rig floor across the entire field without having to touch the ground. Those rigs became soaked in oil, gas constantly

*This 55,000-barrel tank has burned so long that one side has collapsed and it is in the process of boiling over, representing one of the most dangerous stages of an oil tank fire. (Photo courtesy of Oklahoma Historical Society, Devon/Dunning Petroleum Industry Collection)*

escaped to the atmosphere, and machinery, including gas fired boilers, operated around the clock. The place was a fire trap waiting for something to happen. Miraculously it was more than a year before the first conflagration occurred. It began in the Hogg-Swayne division and destroyed more than 150 rigs before it was contained by dynamiting a row of derricks and tumbling them back into the blaze to create a fire lane. Although rig fires were dangerous, unlike the Hogg-Swayne incident, they were usually isolated to one rig and were easily extinguished.[11]

Despite the obvious danger from rig fires the most spectacular blazes almost always resulted from tank fires. They usually began when the storage vessels were struck by lightning. The greatest danger from tank fires lay with a phenomenon called "boil over." This occurred when the heat of the fire reached the water that always collected at the bottom of the tanks. At that point the resulting boiling water caused the oil above it to froth and boil over the top of the vessel and flood the surrounding area. Perhaps the greatest disaster on record involving that situation happened at Humble, Texas, in 1907 when a 55,000 barrel tank burned. Crews of men using teams of mules were dispatched to the location in order to build dirt containment dikes around the burning vessel. These dikes would isolate the burning vessel from nearby tanks. Unfortunately, all involved misjudged the stage to which the fire had progressed and the resulting boilover covered the earthmovers. Bill Bryant, who was one of those who went into the area to recover the bodies, described the situation.

> I think it burned for seven days before we got in there and helped pick up the men, and they was mules and scrapers, and mules and plows all over the tank farm and I believe we picked up seventeen men . . . the only way they ever knew how many men was lost was from the Texas Company payroll, from the books, that didn't show up.[12]

Despite the obvious danger from fire and explosion all those early rigs received illumination from open kerosene lanterns. The most commonly used was known as the "yellow dog" lantern. It was shaped somewhat like a teakettle with a spout at either end. Wicks were placed in the spouts and the reservoir was filled with fuel. At night one of those devices was placed on the floor of the rig and another was hung in the derrick. As drilling depth neared the pay zone and the danger of gas increased, the lamps were extinguished. From that time until the well came in the crew ran on "moonshine." Bill Bryant related that many times he drilled all night without being able to see any of the men in his crew for most of the time. He made one of the roughnecks sit beside him close enough to touch and ran the rig by the "feel" of the clutch. When it was time to make a connection, that is, add a section of drill pipe to the drilling string, or some other chore, he would have that man get the rest of the crew together and they would make the connection in the dark. Under those conditions it was virtually impossible to see if the new joint was lined up straight enough to screw into the drilling string, so they would move it one way and then the other until it started well enough to tighten. They accomplished the feat by feel and yelling to the derrick man to push the top of the pipe toward town or toward the closest rig or toward any visible landmark. A little electricity would have gone a long way in those early days![13]

During the normal course of events drilling crews were not beset with gas blowouts and fires nor did they have to work without light. The normal routine of drilling oil wells progressed from day to day as well diggers learned more and more about their profession and became more and more efficient at their work. They had no geologists to tell them when they were in the "pay zone" where oil was likely to exist. They depended upon the hard-learned lessons of past experience. For example, when drilling on a salt dome they soon learned to recognize the caprock, that formation that sealed the oil-producing zone, by the way the bit reacted as it entered that rock strata. With that warning of approaching the expected depth of finding oil they closely examined the cuttings from the well and even tasted them for the acrid taste of petroleum. When satisfied that they were through the caprock the drillers would change to a smaller bit and drill carefully until they lost the return flow of their drilling mud. At that point they felt they were in the pay zone because the porous rock that produced the oil had probably absorbed the drilling fluid. Then they ceased drilling and began to bail the well. If oil began to flow they had made a well. It usually did.

Actually it was a bit more complicated than that. The more successful drillers learned to recognize the type of rock they were drilling through by the "feel" of the bit through the clutch vibration, how fast they were "making hole," or by some other indication peculiar to the oilfield in which they were working. They double checked that knowledge by regularly checking the drilling fluid discharge in order to find a familiar type of cutting that would indicate their location beneath the earth relative to known oil reserves. Additionally, they watched the drilling fluid discharge in case it began to "rainbow," or have an iridescent hue. If that happened it was a sure sign that they were in a pay zone and that petroleum was discoloring their drilling fluid. All those practical bits of knowledge combined in the drillers' minds to give them the ability to successfully complete a well. It was as much an art as an occupation. A good driller felt his way down through the earth toward awaiting oil and he knew when he was at the right point.[14]

Although the driller used a learned intuition to do his job in an efficient manner, the equipment he used was both inefficient and dangerous by late twentieth-century standards. The boilers that provided the steam power to the operation often utilized streams or other locations that provided contaminated water that sometimes caused steam explosions. The drive chain that drove the rotary table was totally exposed and many a roughneck lost fingers, arms, legs, or was killed when entangled in the machinery. John Alexander described his experiences while working on drilling rigs during the 1919 boom at Ranger.

> When I went to work in Ranger, Texas, the tools and ways of doing things were very primitive with no safety devices or nothing. Them old chains was slamming and clanking all day and breaking and none of them covered up and boilers blowing up. Most of them old cable tool drillers was from West Virginia and Pennsylvania and up there they thought the steam had to be a dry blue steam. That didn't let you have much water in the boiler. When you got busy and was late checking the boiler it might be dry. If you turn water in on it you might have an explosion. I was lucky. I never blew one up.[15]

*The drilling floor on an early rotary rig showing the driller manning the brake. Note the exposed chains and other drive mechanisms as well as the driller's soft hat and lack of steel-toed boots, all of which give some indication of the extreme danger of early day oil well drilling. (Photo courtesy of The Petroleum Museum, Midland, Texas, Abell-Hanger Collection)*

The derrick man, balanced on a two-by-twelve plank high above the rig floor, handled heavy drill pipe without the benefit of a safety belt. Nobody wore hard hats and tools often fell out of the derrick to the detriment of those working below. Inexperienced drillers attempting to pull stuck drill pipe from "downhole" with the powerful hoisting equipment occasionally "pulled the rig in" when undue stress collapsed the entire superstructure around the crew's ears. Those kinds of accidents and many other dangerous on-the-job activities caused roughnecks to get "a lot of rabbit in them." At the first hint of an unfamiliar sound while working they ran first and looked upward later. Given the dangerous nature of the work, experienced roughnecks soon learned to work only for proven drillers if they had any choice in the matter.[16]

As the practical application of drilling knowledge grew, the technical aspects of drilling equipment grew more and more sophisticated. Problems encountered during the course of well drilling were solved on the spot by developing specialized tools. The thousands of nameless drillers and roughnecks who invented most of those objects out of immediate necessity have made a priceless contribution to the industry. This is especially true in the area of fishing tool application.

During the course of oil well drilling the drill pipe is sometimes twisted off, bits are lost in the hole, or objects are accidently dropped into the well bore. In order to continue drilling these objects have to be removed or "fished" out of the hole. Tools like overshot fishing tools, those let down into the well that fit over the lost pipe, or undershot tools, those let down into the well that go inside the lost pipe, soon appeared as devices to remove twisted-off pipe from the well. A driller who twisted off the drill pipe was usually fired on the spot and was always in a lot of trouble with the boss. To "twist off" soon became the oilfield term for doing something that caused trouble. For example, if one of the hands gets drunk and creates problems for himself in today's oilfield he is said to have "twisted off."

A variety of other fishing tools proliferated during those days. A set of fingers, often referred to as junk baskets, were developed to fish small objects from the well. They probably appeared as soon as the first roughneck dropped the first sledgehammer down the hole. The baskets were simply a piece of pipe with fingers several inches long cut into the lower end of the pipe. It was lowered downhole on the end of the drilling string and slowly rotated after the pipe touched bottom. As more and more weight was applied the pressure forced the fingers to curl inward and upward, plucking the lost object from the bottom of the hole. Then the basket was withdrawn from the hole with the offending object in its grasp. That simple tool varied tremendously in construction, and it was individually crafted for the particular problem associated with a particular fishing job. By 1910 or so those types of tools and many others were used on a regular basis and then relegated to the scrap heap as soon as their immediate need passed.

The whipstock was another important invention that miraculously appeared with no apparent author. Still used by the oil business around the world, the whipstock is a simple device used to "sidetrack" or divert the well bore at an angle from its vertical direction. It was first used to drill around an object lost in the well after all fishing attempts at recovery failed, but with time it evolved as the basic tool in directional drilling. That is, it helps put the bottom of the well at a particular horizontal distance from the place on the surface where the well begins. A well-publicized example of that process is the Petunia #1 in Oklahoma City that was completed beneath the state capitol building, but began on the surface more than 400 feet south of the structure. The whipstock is a simple wedge-shaped device ranging in length from eight to fifteen feet. When first developed it was often made of wood but soon began to be manufactured from steel. It is placed in the well with the thin edge facing upward against the opposite side of the bore hole from the direction you wish the hole to go. As the drilling string is lowered into the hole the bit is diverted away from the vertical as it slides along the thicker portion of the whipstock.[17]

One of the most important and perhaps the best-documented inventions of the day was that of the rock bit, or roller cone bit. The type of rotary bit developed at Corsicana and used exclusively during the early Gulf Coast booms was called the fishtail bit. It was a chisel-shaped device whose cutting edge scraped away layers of earth in order to create the hole. Nobody was entirely pleased with its performance, particularly when trying to drill through hard rock formations. Despite the application of various hardening agents to the cutting surface of the fishtail it made pitiful progress through hard formations. Oil people devoted much discussion on how to improve the drill bit. John S. Wynn finally solved the problem. Having successfully worked as a driller at both Corsicana and Spindletop, Wynn established a machine shop and supply house at Sour Lake shortly after the boom hit there. He, like most others in the business, speculated on how to solve the slow drilling problem of the area. One day as he sat in his machine shop idly playing with a device called a pipe setter, he began spinning the tool on the floor and noticed that as the pipe setter rotated on its roller cones it made a perfect eight-inch circle. As he continued to tinker with the device he gradually conceived the idea of making a drill bit with the cutting surface composed of two sharp-edged rotating roller cones that would chip away the surface of the rock instead of slowly scraping it away.[18]

Wynn invented and patented the roller cone bit in 1908, but it did not meet with overnight success. The prototype lay around his Sour Lake shop for months where it was used to drill through marble slabs and other demonstration materials. Later, Wynn moved his business to Batson and the invention accompanied the move. Several technical problems still puzzled the inventor who doubted he would ever make the thing work as well as it should. Finally, a foreman for the Sun Oil Company who was experiencing a particularly difficult drilling problem asked, as a last resort, to borrow the bit. His driller protested, "Oh, there ain't no use in putting this thing on. It ain't gonna make no hole." But the company man allowed that it was not costing anything, so they might as well try it.

After the bit was attached, lowered into the hole, and drilling began, the driller thought that the pipe had twisted off due to the ease with which it turned. The foreman persisted with the experiment and after drilling sixty-five feet the bit began to drag. Upon pulling it from the well they discovered that one of the cones had chipped and locked up. Wynn cleaned and repaired it and they finished drilling through 134 feet of hard rock in record time.[19]

Walter Sharp heard of the invention and sent a business associate, Howard Hughes Sr., to investigate. Hughes borrowed the bit, took it to Houston, and was impressed with its potential. In 1909 Sharp and Hughes paid John S. Wynn $6,000 for the patent. Evidently, there were a number of other attempts afoot in various parts of the country at developing similar devices. Sharp and Hughes bought the rights to several more bits and formed the Sharp-Hughes Tool Company to develop them. Howard Hughes had the engineering ability to refine the invention into a highly efficient drilling device by using the Wynn model as the basic concept. He and Sharp considered the device so important that all their drill bit experiments were conducted with the greatest degree of secrecy. The bit was taken to the drilling rig in a gunny sack where the crew was ordered off the rig while the bit was attached to the drill stem by Hughes or one of his engineers. It was then lowered into the hole before the drilling crew was allowed back on the rig floor. When the bit was withdrawn from the well only the Hughes people handled the device. By 1910, when the final version was perfected, it rapidly became accepted as the ultimate rock drilling tool on rotary rigs.[20]

Thus, during those first ten years of activity along the Texas Gulf Coast a group of independent-minded hard-working young men developed the basis of oil well drilling traditions that became the norm in the industry. Being young and coming from an agricultural background where hard work and long hours were expected, the ability to endure the backbreaking toil of well drilling seemed routine to them. Those able to excel physically were the heroes of the day. Additionally, the ability to improvise, another necessity of a farm background, fitted perfectly with the needs of oil well drilling. That characteristic also became a much-desired quality in those who followed the oil well drilling trade. So it was that the youth who flocked to the early oilfields developed the ideal of men who endured hardships, resented close supervision, and used a native ingenuity to develop a respected trade where none existed before.

Although born at Corsicana and brought to maturity on the Gulf Coast, rotary drilling did not stand alone as "the" drilling process. Most cable tool drillers refused to become a part of the upstart technology. After all, cable tools had stood the test of time since the Drake well in Pennsylvania more than forty years earlier. They maintained that their method, which one cable tool driller characterized as "like stomping a hole in the ground," had distinct advantages over rotary work. The most important objection dyed-in-the-wool cable tool men gave to rotary drilling dealt with the primary purpose of finding oil. They argued that their slower drilling speed allowed them to identify oil-producing zones that the rotaries often zipped past. Further, they contended that the mud used in rotary drilling created a clay sheath that tended to seal off any porous low gas pressure production zones as they were drilled through, which prevented the finding of potentially good production. The stage was set for a strong competition to develop between the traditionalists and the innovators.[21]

That competition developed into a strong animosity during the first years of the Gulf Coast activity. As rotary man Dan Lively stated, "I've never seen anything around one of those cable tool rigs that a mule couldn't do if he had two hands."[22] Because so few experienced cable tool drillers converted to the new rotary technology, that new skill soon became dominated by the throngs of young men new to the oilfield. The success those rotary drillers experienced created an intense pride that soon blossomed into an overwhelming sense of

*The drilling floor on an early cable tool rig, showing the driller standing on the drilling stool while monitoring the tension on the drilling line and the "toolie" leaning against the "headache post" with his foot touching the "circle jack." (Photo courtesy of The Petroleum Museum, Midland, Texas, Abell-Hanger Collection)*

self-importance. As one cable tool driller explained, "Now those rotary drillers, damn them, ever one of them thinks he's God almighty."[23] There was even the apocryphal story circulated widely among the oilfield hands of two cable tool men walking down the street one day when they happened to spy a sign for the local Rotary Club. One of them turned to the other and declared in disgust, "The sons of bitches even have a club."[24]

Cable tool men soon began to refer to rotary people as "swivel necks" or "mud eaters," and they were, in turn, called "rope chokers" or "jar heads" by the rotary men. At every opportunity they would get into fistfights and each group tended to congregate at their special bars or cafes and not mingle socially if at all possible. By the time of the Electra boom in 1910 one rooming house proprietor discovered, much to his dismay, that renting rooms to both groups had a tendency to get the establishment wrecked when the principals got into a discussion on the relative merits of their particular specialties.[25]

The reality was, that although the rotary came into prominence as the only technique that could successfully drill in the soft Gulf Coast formations, the cable tool outperformed the rotary in all the harder formations scattered across

*The driller and his "toolie" pose holding sixteen-pound sledgehammers in readiness to dress (sharpen) the drill bit lying between them, ca. 1910. Note the driller wears waist pants while the "toolie" is dressed in overalls as an indicator of the relative importance of their status on the job. (Photo courtesy of Oklahoma Historical Society, Ira M. Spangler Collection)*

the state. The concept of "mudding off" a low pressure producing zone was also accepted by many of the Gulf Coast producers. Consequently, the oil companies often drilled their wells close to the production zone and completed their wells with a cable tool under the presumption that the older method would assure them of not missing the pay zone. Naturally, bringing the contentious working groups into close contact did little to stop their competition.

Despite the competition between the two groups, their working situations were very similar. Whereas a rotary rig required four roughnecks and a driller, the cable tool operation needed only a driller and a tool dresser, which made a cable tool operation much less expensive to operate. It was generally conceded that it took five years of experience to develop a good cable tool driller, after which he, much like his rotary counterpart, assumed the role of supreme arbitrator over how the rig was run. His tool dresser, or "toolie," performed most of the subsidiary tasks associated with keeping the rig operating, such as tending the boiler and sharpening the bits, from whence his title of tool dresser derived. Both types of crews worked a twelve-hour shift or tour, pronounced "tower," that extended from noon to midnight. There has never been a satisfactory explanation of how tour became tower, but Gerald Lynch, who spent thirty-five years in the oil patch, explains it better than most. He claims that because most of the hands were so poorly educated, they tended to spell phonetically. The word *tour,* which was unfamiliar to them, was likened to *sour,* which was familiar to them, and tour then became tower.[26]

Thus, during the first years of the twentieth century the nature of the drilling crews in Texas became firmly established. Cable tool operations were definitely accepted as the norm with their forty years of proven experience dating back to the 1859 Drake well in Pennsylvania. Rotary drilling was in its infancy. It was considered by most to have little practical use beyond the specialized needs of the Gulf Coast and other areas where soft unconsolidated formations existed. That was to remain the nature of drilling technology during the first twenty years of the twentieth century as the vast majority of the wells were completed at less than the four-thousand-foot level and the need for significant technological change remained low.

## CHAPTER 3

# The Other Hands, 1901–1910

Although well drilling receives the most notice of all occupations in the oil patch, many other workers combine their efforts to make the industry operate. Those jobs range from the exotic, like well shooters and oil well firefighters, to the less glamorous, such as pipeliners, teamsters, roustabouts, tank builders, and a host of lesser-known occupations. It was the influx of all those other hands to support the frenzied well-drilling activities that caused the boomtowns to boom. Once the flush production ended and the intense pace of drilling slowed, most of the drilling crews as well as the vast number of support personnel necessary to open a new oilfield moved on and the boom ended. It was rare for the hurly burly of oil boom activity to last longer than one to three years at any given location. As the drilling activity slowed the affected region lapsed into the more sedate lifestyle of producing the completed wells and gradually drilling up marginal areas missed during the excitement of the boom. But once an oil boom touched a region it was never quite the same as before.[1]

Among the first workers to arrive at the scene of a boom and the last to leave were the teamsters. It stands to reason that someone had to transport the heavy equipment in to start the boom, keep the operation supplied, and haul the heavy equipment away when the boom ended. The teamsters did exactly that. During the first twenty years of the twentieth century, long-distance freighting in Texas was dominated by railroads, while animal-powered transport was utilized for short-distance hauling. That type of transportation demanded that workers live in close proximity to the oil discoveries, usually walking distance, which contributed to the chaotic pattern of boomtowns developing in the midst of every big discovery. If the town happened to be firmly established before the boom it usually continued to exist in a much diminished state after the boom, but tremendous numbers of those new entities spawned by the booms simply disappeared off the landscape shortly after the boom subsided.

The Spindletop discovery caught the Gulf Coast region unprepared for the massive freighting demands created by the feverish drilling activity. The first hauling in those oilfields was done with ordinary farm wagons pulled by four horse teams. The lightweight drilling equipment used at the time could usually be adequately handled by those vehicles. Some of them even had the narrow-tired "butcher knife" wheels, although most used the wider tire preferred for the boggy coastal terrain. Hundreds of those light wagons hauled tons of drilling equipment, thousands of feet of pipe, and an untold quantity of general supplies to the hill on a twenty-four-hour basis. Perhaps the busiest place in Beaumont was the railroad yards, which were jammed with hundreds of teams and wagons feverishly unloading thousands of railroad cars crowded on numerous newly built sidings.

Supplying foodstuffs was an important part of the freighting done at Spindletop. Hiram Sloop, whose father operated a grocery store at the hill, remembered leaving home before daylight and driving one wagon while his father drove another. They went into Beaumont where they loaded up with meat, canned goods, and a variety of other supplies for their store. They usually arrived back home by 6:00 A.M. and prepared for the day's business. Young Sloop's job at that point was to deliver gasoline to the rigs and other business establishments on and around the hill using a light one-horse delivery wagon. Each day the store received between fifty and one hundred of the fifty-five-gallon drums of gasoline from a local refinery. Also approximately four hundred empty gasoline-fired lamps were brought there daily by local businessmen to be refueled from the drums. Sloop's job was to fill the empty lamps and return them to their owners. During the delivery process he also refilled those lighting devices still in the establishments that were not totally empty. Early in the afternoon the two big Studebaker wagons went back to Beaumont for another load of staples. During that second trip of the day most of the supplies were delivered directly to the boardinghouses in the area, which many times took an entire wagon load directly from the wholesale house. Considering the volume of business from the Sloop store, the existence of several other grocery stores around the hill, and the fact that many of the boardinghouses and restaurants hauled their own supplies, the volume of freight traffic in food supplies alone was tremendous.

With all that activity and the significant amount of rainfall in the region roads soon developed into almost impassable quagmires. When the main thoroughfare to the scene of the action at Spindletop became too muddy to use, the drivers simply detoured out into the fields and bypassed particularly boggy places. As this activity continued to escalate the road became a hundred or more yards wide in places. During those first months of the boom local horses and mules were used to pull the wagons, but within a year larger and stronger draft animals were introduced from the East to better cope with the heavy hauling and the poor road conditions.[2]

Despite the magnitude of the generic freight traffic during the Gulf Coast booms it was dwarfed by the massive freighting efforts necessary to supply oil well drilling and production needs. When the Hamill boys arrived in the fall of 1900 they hauled their drilling rig and associated equipment in a lightweight Studebaker farm wagon. However, as the boom escalated and the giant piles of pipe, rig timbers, boilers, and drilling machinery accumulated in the rail yards larger freight wagons pulled by bigger teams of horses began to be used. By the end of the first year of the Spindletop boom eight-wheeled freight wagons with ten-inch-wide tires were in common use for the heavier loads, although lighter wagons using a four-horse hitch continued to haul the bulk of the general freight.

Within a couple of years, as the Sour Lake and Batson fields were discovered, an entirely new problem cropped up. How to get the same types of supplies to the new fields across twenty or more miles of boggy coastal plain created particularly difficult problems. The topography consisted of a generally flat stretch of land interrupted only by occasional streams or bayous along which rows of trees grew. Low ridges that usually remained dry crossed the area in various places; all the lower-lying land was dotted with ponds filled with water during all seasons of the year. Freighters in this typical coastal plain environment traditionally used the ridges as roadways because of their dry firm nature and avoided the low-lying areas except for the ponds, which they utilized to water their teams. This feat was normally accomplished by actually driving the draft animals into the ponds and allowing them to drink without removing their harness. It seems that most of the ponds had hard sandy bottoms capable of supporting the weight of all except the most heavily laden wagons.

Both Plummer Barfield and William Cotton worked as teamsters in those days. They usually drove teams of four up, that is, four horses abreast, or two four ups if the load was particularly heavy. Horses or mules were ordinarily used and they were guided by reins and a lot of imaginative swearing. However, in particularly boggy situations oxen were preferred as draft animals. Their cloven hooves gave them extraordinary traction in the slippery mud, and they also pulled in a slower and steadier manner than their equine counterparts. The oxen were guided by voice commands and by the use of long bullwhips, which were more often used to tap on the animals' horns to indicate direction rather than to inflict pain. It also took a considerable amount of swearing on the part of the teamsters to drive oxen in the traditionally accepted manner.[3]

Landon Cullum recalled a particular incident illustrating freighting problems in the Gulf Coast region. In 1910 or 1911 he was in charge of construction on a pipeline pump station some three or four miles from Sour Lake. Fortunately, the railroad had arrived in Sour Lake a few years earlier so the steam boilers he needed were delivered to the train depot without incident. However, it was a particularly wet year and the boilers had to be transported the rest of the way to the pump station utilizing animal-drawn vehicles. After numerous attempts it became obvious that neither horses nor mules were able to budge the heavy load in the deep mud. Finally, they loaded the boilers on sleds constructed from rig timbers and hooked ox teams to the sleds. Eventually, sixteen yokes of oxen hooked to each side of the sled, a total of sixty-four animals, labored for two days to move each of the boilers the three miles or so to the construction site. Working on the muddy coastal plain was indeed a tedious and nerve-wracking job during the early days of the twentieth century.[4]

*Oilfield teamsters at Spindletop, ca. 1905, with an "eight up" team of horses hitched to a heavy-duty freight wagon sporting wide tires. (Photo courtesy of Southwest Collection/ Special Collections Library/Texas Tech University, C. C. Rister Collection)*

Although almost all the oilfield jobs during those days were reserved exclusively for whites, freighting was one of the few occupations made available to the African American community. Although black men were responsible for transporting a variety of freight, African American teamsters are best remembered as doing most of the large-scale dirt excavation work in the region. Using mule teams and Fresno scrapers they dug various types of disposal pits and thousands of dirt tanks used for oil storage vessels. The African American workers were always quartered in segregated areas away from the main oilfield activity. At Batson, black teamsters lived about three miles away on Pine Island Bayou. They also worked their teams differently. Contemporaries remember African Americans singing to their animals, a practice totally foreign to white teamsters. Late in the afternoon when both men and animals grew weary from their labor and were hungry for their supper a colorful scene developed.[5]

> As a rule, you know, a Negro, he always sung everything he said, along before sundown especially. And in the big camps, in the grading camps there were lots of Negroes and lots of mules, why, then the Negroes would go to singing, why, the mules would go to braying because it was pretty near to sundown. They made up their songs as they went and pretty muchly some of them was in the same category that they sing in the spirituals right now.[6]

One of the favorite sayings of the oilfield hands to indicate how long they had been in the business is that they rode into such and such a boom on the first load of rig timbers and they rode out on the last load of drill pipe. That pretty much sums up the job of the teamsters. They were the first ones in when they brought the drilling equipment and they were the last ones out when they hauled off the unused pipe. Hard on the heels of the teamsters came the rig builders who erected the derricks and installed the drilling equipment so the work at hand could commence.[7]

As the experience of the Hamill brothers illustrates, when the Gulf Coast boom began there were few, if any, specialized rig builders in Texas. The work of building the derricks and installing the drilling equipment was done by the drilling crews. This combination of derrick and drilling equipment constituted a drilling rig. The job, which took about a week, was accomplished much like how the Hamill boys built their rig in 1900. The rig timbers were "laid out" on the ground in the desired conformation and cut into the needed length. The slope, or "batter," of the derrick was determined by its height and base dimensions as envisioned by the person in charge of construction. Consequently, there were no standard rig sizes and shapes in those first days due to the varying ideas of different bosses on different jobs.[8]

Most of the derricks built in Texas between 1900 and 1915 stood around eighty feet tall, plus or minus 10 percent, and had a base approximately twenty feet square. All of them were constructed of wood and their bases began with timbers or sills eighteen to twenty-four inches square by twenty or so feet long. Above this foundation six-inch by six-inch timbers were laid in a pattern two feet apart and crosswise to the sills. Then a layer of two-by-twelves was nailed across the six-by-sixes to make the floor of the rig. At that point the derrick legs, made of two-by-twelves, nailed in an L shape, were fastened in an upright position on each corner of the floor. The proper batter of the derrick legs was determined by two-by-twelve girders, or girts, nailed horizontally at approximately seven-foot intervals up the length of the derrick. Additional bracing in the form of two-by-eights, called sway braces, were nailed in an X pattern between the girts. The entire structure was topped by a "crown block," which contained sheaves or pulleys through which the drilling cable was threaded. The platform, high atop the derrick, that supported the crown block assumed the name of the "water table." The entire derrick, except the sheaves proper, was constructed of wood.[9]

As the boom area expanded the new discoveries made the building of hundreds if not thousands of derricks necessary, and a new specialization developed. Soon certain individuals who showed an aptitude for rig building began to do only that work. By 1904 or 1905 most of the rig building on the Gulf Coast was being done by specialized crews of five or six men who could produce a rig ready to drill in about five days. Along with this specialization of labor there also developed a standardization of derrick sizes and the associated drilling machinery. Supply company catalogs specified the exact numbers and sizes of all the lumber items needed for a particular-sized rig. Using that standard pattern all the wooden elements could be precut at the lumberyard and delivered to the drill site ready to be assembled. Thus the speed and efficiency of rig building was greatly enhanced.[10]

Rig builders were renowned throughout the oil patch as a special breed of worker. The labor was not exceptionally harder than most of the other work in the industry, but what eliminated most hands from the occupation was the fact that much of the activity took place high above the ground, making it a particularly dangerous line of endeavor. Consequently, rig builders drew wages that were usually double that of ordinary workers. They earned their good wages because, as is the case with most semiskilled labor that pays well above average, what they did was either so difficult or so dangerous that most ordinary men shied away from it.

A rig builder's primary tool was his rig ax. The rig ax was an exceptionally heavy hatchet with a hammer head opposite the cutting blade. Using only that tool, a hand saw, and forty penny nails called rig spikes he built his rig. After standing up the first course of rig timbers and nailing the first row of girts nine feet above the rig floor the rig builders put up a single two-by-twelve plank across the girts to act as a scaffold and nailed in the next two-by-twelve girder seven feet above the first. This was done by two men holding the girt above their heads with one hand and utilizing their rig axes to nail it in place with the other hand. Once all four girts were in place for that particular course they nailed in the two-by-eight sway braces to give the structure stability. If for some reason a particular piece did not fit properly the rig builder simply reversed his ax to utilize the blade side and modified the size of the offending board on the spot and then nailed it into place. Once a course was in place the workers raised their single-board scaffold and added another section of derrick leg and repeated the process. As the derrick grew higher three derrick hands nailed the thing together while two ground hands pulled the timbers to the air crew using a single sheave rope and pulley apparatus. By working nine and ten hours per day a typical five-man rig-building crew was able to complete a derrick in less than a week.[11]

By around 1920 this construction process was so refined that it became possible to do the job in three or four days depending upon how accessible the work location was to the transportation of the day. Although building a complete rig in twenty-four hours was unheard of, at the time of the Burkburnett boom in 1918 a rig-building contractor named O. W. West overcame that obstacle. It happened when he was contacted by a Kansas City oil promoter who had overlooked his spudding in deadline on a lease in the Burkburnett area and would lose his lease if he did not begin drilling on a date specific. The promoter talked to West on a Friday night and declared that he had to have his well begun by midnight on Monday. The man begged pitifully for West to help him out of his dilemma.

*Rigging up a cable tool rig on location in 1914 in preparation to begin drilling. The rig builders are hoisting a "girt" to complete the next-to-last section of the derrick. The "bull wheels" are on the ground beneath the derrick and the "band wheel" is on the ground behind the buggy, both of them awaiting installation. Meanwhile, a worker is hooking up the steam lines to the boiler. (Photo courtesy Oklahoma Historical Society, Devon/Dunning Petroleum Industry Collection)*

The rig-building contractor finally relented and agreed to do the job in the time allotted in return for payment on a double or nothing basis. Accordingly, West and his crew finished the contract with time enough to spare for the man to install his equipment and begin drilling before the deadline expired. After that West advertised that he would put up a rig anywhere in the field on twenty-four hour notice if the drilling contractor had to meet a deadline.[12]

As with many of the early day oilfield occupations rig building was dominated by hard-bitten West Virginians whose cantankerous exploits became legendary in the patch. One of those contrary old hands was still working as late as 1926 during the Borger boom. He was particularly cranky about his work clothes. Actually, when he bought new overalls he never washed them but simply wore the apparel until they became so ragged and dirty that he had to replace them. The part he was so fussy about was the way he altered the overalls so that the pant leg length was exactly even with the tops of his work boots. He maintained that this kept the cuff of the legs from snagging on something that might cause him to fall out of the derrick. One day he took a newly purchased pair of overalls out on the job and carefully folded the legs to the exact spot he wanted them cut off. While he was searching the location for a rig ax to cut the legs to the proper length, one of the young hands in the crew refolded the clothing. The old West Virginian soon returned and in two or three swift blows altered the apparel. Upon donning his new work duds he discovered that he was the proud owner of knee-length britches. For the next hour or so he brandished the hatchet, turned the air blue with profanity, and swore to kill the offender. However, nobody in the crew was able to remember anyone tampering with the overalls so he was reduced to fussing and fuming for the rest of the day.[13]

Once the rig builders finished their work those lords of the oil patch, the drillers, took over. They drilled their wells and had their problems as one might expect. Despite the drillers' ingenuity they occasionally encountered obstacles they could not overcome. One of those obstacles was blowouts. Blowouts were the result of drilling into unexpected pockets of high-pressure gas that wreaked havoc on the drilling equipment and sometimes the crews of drilling rigs. That dreaded event occurred on the Spindletop discovery well, and later in the Gulf Coast field of Humble it became a particularly difficult problem when drillers encountered unexpected gas pockets that often got out of control. Those blowouts sometimes caught fire and if they were not quickly controlled became labeled as wild wells. If those wild wells continued to blow for an extended period of time a situation known as cratering tended to develop. The phenomenon occurred when the extremely high pressure gas and fluid blowing from the wellhead widened the hole to create a crater that could extend a hundred feet or more in diameter. If this happened there was a real danger that the resulting hole could swallow the derrick and all associated drilling equipment. Bill Bryant remembered one such event and the subsequent results.

> I lost the first rig to the Texas Company that was lost in Humble—on the Moonshine #28. Johnny Lynn was running nights and it was jest as we changed tours the well started blowing out and an hour from the time it started we didn't have anything left but our boilers. Everything had fell into the hole, the crater. I was pretty blue about it, losing the rig, and about three days later, Mr. Beatty, he lost one. And Mr. Lyons at that time was our pusher. Well, I naturally thought that Mr. Lyons would come down all blowed up, lost the whole drilling rig, derrick and everything—and first thing he asked me, he said, "Billy did you lose any men?" I said, "I never got one hurt." He said, "You was lucky." And that was all he said. He was a good old man.[14]

Naturally, putting out well fires before they cratered and created more severe problems assumed top priority for the oil companies. Usually, existing steam boilers, augmented by others brought in for fire-fighting purposes, allowed the drilling crew to smother the blaze with jets of steam before it got out of control. If that technique did not work a variety of schemes were tried from time to time. They even included such far-fetched ideas as shooting the flames out with military cannons. On one occasion the Sharp brothers excavated a one-hundred-foot-long tunnel about ten feet beneath the surface, cut the casing on a wild well, and installed a valving arrangement that shut off the fuel to their burning well. Ultimately, however, the most successful way to extinguish burning oil well fires became known as shooting. Shooting out well fires created a small group of specialists in perhaps the most dangerous occupation in the oil business.

C. M. Chester is given credit with shooting out the first oil well fire in Texas. In 1908, Chester, a superintendent with the Texas Company, had a particularly difficult well fire near the Louisiana state line. The derrick and all the drilling equipment were destroyed in the conflagration and flames were shooting 175 feet into the air. The usual method of using steam to smother the fire had failed and Chester was desperate. Then he came up with the idea of extinguishing the blaze with a dynamite blast. Accordingly, he built a steel cage to hold twelve sticks of explosives wrapped in a covering of asbestos. Then he stretched a small cable so that it was connected to pulleys on either side of the fire and attached the cage to the cable. After that he maneuvered the explosive-laden cage directly over the wellhead and exploded the device. The resulting explosion was great enough to temporarily deprive all the oxygen from the burning area and extinguish the flame. That basic process, with the modification of substituting nitroglycerin for dynamite, has served to the present day as the last resort method of extinguishing out of control oil well fires.[15]

Shooting out oil well fires was actually an outgrowth of one of the more exotic of the oilfield occupations, that of the shooter. Shooters were those workers who lowered nitroglycerin torpedoes into the wells and exploded them in order to fracture the rock and expedite the flow of oil into the well bore. Although the practice had existed since the first oil discoveries in Pennsylvania, it was seldom used in the soft Gulf Coast formations where the oil flowed freely. However, by 1910 it was a common practice in other areas of the state where wells were shot with nitro in order to shatter the dense stone of known producing formations in order to release the oil trapped within the fine-grained rock. It took an unusual person to follow a line of work that required the rare combination of a man who both understood explosives and who had the nerves of steel necessary for the job.[16]

There are scores of examples of the dangerous exploits of those men. One that surfaces time after time with different characters illustrates what it takes to do the job. Ed Matteson related how he completed a well near Breckenridge sometime in 1919 or 1920. A shooter by the name of Davidson with the Independent Torpedo Company went out to do the shooting. That particular well had been doing what the hands called *breathing.* Every forty-five to sixty minutes

*A shooter carefully pours a can of nitroglycerin into a shell suspended over the well bore, ca. 1920. Note the witch elk lace-up boots he wears; these were the preferred footwear for oil-field hands of the 1920s, providing they could afford them. (Photo courtesy The Petroleum Museum, Midland, Texas, Abel-Hanger Collection)*

low-pressure gas would build up and cause a small amount of oil to flow out of the well. The shooter waited for about three hours for the event to transpire again and when it failed to happen he decided it was safe to begin his work. Assuming the well was dead he filled one of his torpedoes or shells with thirty quarts of nitroglycerin and began to lower it downhole. Before the device reached the bottom the well came back to life and the nitro-filled shell could be heard scraping the sides of the casing as gas pressure forced it upward. Matteson and his toolie left the rig floor at a dead run. They did not stop until they reached a deep ditch about one hundred yards away. After diving headlong into the protective cover they waited anxiously for the explosion sure to occur as the nitro emerged from the well. The sound of oil hitting the crown block rang in their ears as both men timidly peeked over the edge of their hiding place. Then they stared in amazement as the shooter put his hand into the oil flow, caught the bail of the shell as it emerged, and calmly lifted the thirty-quart container out and deposited it safely against the rig timbers where it could not fall over. Shortly afterward, as the shooter showed the drilling crew the hand he had cut while performing the deed, his hands showed not the least bit of a tremor. That type of reaction is what set shooters apart as a different breed.[17]

Ultimately, there were many specialists in the oil patch. Naturally, workers like rig builders, shooters, and firefighters received lots of press due to the dangerous nature of their work, but others abounded. Roustabouts were perhaps the most numerous of the working-class hands. Working in a roustabout gang was where most oilfield workers began their careers because it required little if any specialized knowledge beyond being able bodied. Roustabouts did all the general cleanup and maintenance in the oil patch. They maintained the wells after they were completed, laid most of the small-sized pipelines from the wells to the tank batteries, and generally kept the area neat with painting, fencing, and small repairs. At various times roustabouts were called either bull gangs or connection crews, and they were always led by a pusher or gang pusher. Working successfully in a roustabout crew many times determined whether one remained in the oil patch or moved on to some type of less-demanding line of work.[18]

Another specialized but lesser-known oilfield job is that of casing the well. For a variety of reasons ranging from sealing off water formations to supporting the sides of the well bore it is necessary to insert a pipe lining into the well bore. For example, one might case off the first three hundred or so feet of a well where water-bearing formations lie. Another section of the well where there is an unconsolidated formation might also require casing. Each of these casings extends from a given depth in the well to the surface. For example, if a well contains three runs of casing when one looks down at the top of the well on the surface one would see three concentric rings of pipe. As one might imagine, putting all that casing into a well in twenty-foot lengths of pipe requires a considerable amount of labor.[19]

In the first days of the Gulf Coast boom drilling crews always cased their own wells. But very quickly the use of specialized casing crews developed throughout the oil patch. By 1919 those crews were operating on a regular basis throughout the Texas fields. Casing crews were able to do the job faster and more inexpensively than the drilling crews because that was their sole occupation and they became very efficient at the work. At a time when wages averaged two to three dollars per day the casing crews received five dollars per man for each well they cased. The job could usually be done in one day so the labor meant big money to the working hands of the time. In those days the casing crew, usually five men, tightened the casing by hand using a method whereby a short section of chain was attached to the center of a wooden handle approximately six feet in length. The chain was wrapped around the casing two or three times and four men, two on either side of the wellhead, grabbed the handles and walked in a circle in order to tighten the threaded casing pipe. Once again, as in many oilfield operations,

*Casing crew tightening a joint of casing in 1923 before the days when engine-powered mechanical devices were available. (Photo courtesy Southwest Collection/Special Collections Library/Texas Tech University, C. C. Rister Collection)*

the pusher was the only hand necessary to be knowledgeable, but most casing crews stuck together as a team because it was necessary to work together in a well-coordinated effort in order to complete their work in the fastest time possible to make the good wages they commanded.[20]

The perception has always existed that most oilfield work revolves around well drilling and its associated activities. However, that is far from true. Most oilfield labor is associated with what happens after the well is drilled. The oil has to be stored and transported after it is produced, and that activity requires massive numbers of workers for both the construction and the maintenance of those types of activities.

Oil storage capacity developed as a major problem along with the first well drilled at Spindletop. The tremendous amount of oil produced by those early Texas wells, many estimated as flowing in the 100,000 barrel per day range, created impossible storage problems. In the case of the Lucas well at Spindletop in 1901 workers threw up temporary dirt dikes to contain the massive amounts of oil spewing from the well.[21] Indeed, the earthen pit method of oil storage at the well site remained a common practice in the Gulf Coast region for a number of years. Those temporary earthen containers, still called tanks in the Texas ranching country where the pits store water for livestock, required only earthmoving crews for their construction. That, in part, explains the large number of grading camps in the area and it also represented one of the few occupations open to African American labor in the oilfields of that era.[22]

A variation on the earthen pit storage utilized similar pits enclosed with wooden walls held in place by iron hoops. Those wooden-sided vessels demanded some expertise in the art of tank building. At first they ranged in the 50,000 to 100,000 barrel capacity, but within a few years the Gulf Coast became inundated with open storage pits holding hundreds of thousands barrels of oil each.[23]

Tankage of a more permanent nature soon became an urgent necessity. Within the first year of the Spindletop discovery J. S. Cullinan had two 40,000 barrel riveted iron storage tanks built to fill tanker ships coming in from the Gulf of Mexico and within a very short time similar storage capacity to support the activities of the refineries being built at Port Arthur demanded immediate attention. Large metal storage tanks ranging in the capacity of 40,000 to 180,000 barrels were soon under construction in clusters called tank farms. All the tank builders who worked on those jobs were experienced men imported from Pennsylvania where a number of established companies contracted that type of construction.[24]

Those workers were considered some of the toughest in the oil patch and soon added to that reputation in the beer joints and dives of the area. They always seemed to be lumped into the category of boomers, those who only made the boom and then moved on to the next big find, as some of the less desirable elements in the industry. Perhaps the nature of their labor contributed to that reputation. The tanks they built were composed of thick sheets of iron held together with iron rivets. They fastened the sheets together by heating the rivets red hot and hammering them flat with hand tools. The unbelievable hard labor involved in that sort of construction attracted physically imposing specimens. They also tended to carouse after their grueling work days. Many other workers suffered at their hands at one time or another when the tankies were playing in the joints.[25]

In addition to the large earthen pit tanks and the big riveted vessels that make up the tank farms, smaller immediate storage was needed at the well sites. Those tanks, ranging in size from 100 to 1,200 barrels in capacity were of wooden construction. In the beginning those tanks were made of pine or any other particular wood readily available, but they soon became almost exclusively constructed of the cypress that was in abundant supply in the coastal region and had the advantage of better withstanding decay than most other varieties of lumber. Gradually the cypress tanks gave way to the redwood that had many of the same characteristics of permanence but was in more plentiful supply. By the time of World War I most of the small wooden tanks had been replaced by bolted steel vessels, except in areas of high corrosion where wood tanks continued as the best

*A tank-building crew decks a 55,000-barrel steel-riveted storage tank at a time (ca. 1910) when the red hot rivets were pounded tight by hand. (Photo courtesy Oklahoma Historical Society, Devon/Dunning Petroleum Industry Collection)*

product available to withstand the unique corrosion problems created by petroleum products. Regardless of the type of construction, wood, steel, or riveted, tankies continued as a very specialized occupation whose practitioners held a reputation for recklessness among the roughest elements in the oil patch.[26]

By the time a few wells came in on the coast and some storage was available, transporting the oil became a major concern. Railroad cars were pressed into service from the very beginning and some small pipelines took the oil to the coast for ships to transport to the East Coast, but the immense quantity of the oil produced demanded that a better transport be developed. That need was answered by the construction of major pipelines that took the oil to refining areas. As a large number of refineries began to be built on the Gulf Coastal Plain pipelines from other regions as far away as Oklahoma brought oil to the area. That need attracted one of the most interesting sets of workers to grace the industry, the pipeliners. Experienced pipeliners, known in the trade as cats, descended upon a boom in a horde like a plague of locusts and in a matter of days disappeared like a puff of smoke as their job location moved away from the scene of the boom. They were the gypsies of the oil patch.[27]

As one old cat explained it:

> There's a saying in the oilfield that when a man can't do nothing else, he goes to pipelining. And there's a lot of truth in that saying. Pipelining

> ain't no picnic. Even if I do sound like I'm bragging, it takes a real man to be a good pipeline cat. The work will kill the average man and lots of men have tried pipelining, but only for a spell. They can't take it. Digging ditch all day ain't no snap in itself, but us cats dig ditch to rest between pipeline jobs. Yeh, a man goes pipelining when he can't do nothing else, but he don't last a helluva long time.[28]

A major pipeline operation in the 1900–1920 era involved five separate operations. Those activities consisted of the survey and brush clearing, pipe stringing, line laying, ditching, and backfilling/cleanup. Each of those activities was labor intensive and utilized draft animals for all the hauling procedures. In those days the pipelines used threaded pipe ranging in dimension of four inches to eight inches. The pipe was put into place and screwed together by hand. Additionally, all associated excavation was also accomplished by hand. Crews performing the labor numbered in the hundreds and they seldom remained in the same spot for more than a week or ten days. It was a very mobile operation that required a great deal of coordination to make it perform efficiently.[29]

Burt Hull, who surveyed the six-hundred-mile-long Texas Company line from the Glenn Pool oilfield in Oklahoma near Tulsa to Humble, Texas, near Houston, provides an excellent narrative of the work of that early era. His 1906 survey crossed a sparsely settled portion of Texas and Oklahoma whose small towns boasted few boardinghouses or restaurants. The infrequent roads there, always in poor repair, seldom went in the direction desired by the survey crew. Consequently, the survey personnel operated from mobile camps that moved constantly as the survey progressed.

Hull remembered hiring Bill Fondrell of Waxahachie, Texas, who provided two wagons and two teams to serve as camp boss. His job was to set up the camp and move it to a new location on a daily basis. Each morning as the crew left to do their work the camp contractor would strike the tents and move the camp ahead of the surveyors. By noon the camp boss usually had the new camp established and a meal prepared, which was taken back to the survey crew who were notified of the new camp's location. By dark when the survey workers arrived at the new camp a hot meal was ready, and after supper they bedded down to start the process over again at daylight the next day. All went well until they entered the rough mountainous terrain between Coalgate, Oklahoma, and Sherman, Texas, that held no roads or highways of any kind. At noon one day Bill failed to report in and the surveyors were left on their own. It took the crew more than a week to work their way from where Bill and the camp gear disappeared to Dallas. During that entire time they were forced to stay at isolated farmhouses where they slept in barns and depended upon the farmers' generosity for their food. From Dallas to Humble the region was more settled in nature and the crew managed to get by without a camp more easily. About two weeks afterward, with the survey crew operating somewhere between Corsicana and Humble, Bill finally arrived at the Dallas office of the Texas Company with the wagons and the camp gear. Hull fired the erstwhile camp boss on the spot with the remark that "if he's going to get lost, why, he's no good to us."

Hiring good camp bosses aside, the problems of surveying early day Texas pipelines were legion. The first difficulty developed in laying out the route. All pipelines are constructed in as straight a line as possible without regard to topography because variations in elevation interfere much less with pumping liquids from one location to another than creating numerous twists and turns. However, in 1906 developing that straight line proved difficult due to the poor quality of the county maps whose scales varied from map to map. This made it extremely difficult to accurately match up one county to another in order to keep the pipeline right-of-way on course. Consequently, much of the pipeline survey work proved to be educated guesswork. Usually those problems tended to resolve themselves as the landowners realized they would receive between ten and twenty-five cents per rod for allowing the line to cross their property as well as a contractual agreement that the company would maintain the land in good condition.

Burt Hull, in his forty-three years with the Texas Company, could not remember a single land condemnation suit going to court to settle a right-of-way dispute. Nevertheless, he did recall many cases where dogs were sicced on crews who entered some farmer's cornfield or other places where they had not sought out the proper permission. One particular illustrative incident occurred near Conroe, Texas, on the Glenn Pool to Humble project. Late one Sunday afternoon the survey crew stumbled upon an isolated farmstead whose owner was absent. Being hot and sweaty, they took down the farmer's wash tubs and proceeded to fill them with water from the well and take baths in the front yard. Then they milked the cows, gathered some eggs, killed and cooked three or four chickens, and generally made themselves at home. About ten o'clock that evening the farmer and his family returned to find a group of strangers bathing in their front yard and generally enjoying unsolicited hospitality. The next morning, after the farmer got into a better humor, Hull straightened out the affair by paying for the confiscated produce and hiring the fellow to help clear brush from the survey right-of-way. The farmer, an African American, developed into such a good hand that they hired him to continue the work until they finished the survey. As it developed he was the only black person of that era ever to work on a survey crew for the Texas Company.[30]

With the survey completed, work of actually laying the pipeline began. It took six months to complete the eight-inch line from Glenn Pool to Humble. The first order of business was to string the pipe. In that particular instance stringing, the process of getting the pipe laid out on the surface within the right-of-way, proved to be fairly easy because much of the route paralleled the Trinity and Brazos Valley Railroad. The company simply shipped the pipe by rail to various convenient points along the railroad where it was unloaded. Then they contracted with local teamsters to string the pipe out along the surveyed right-of-way where it was unloaded at an hourly rate for the team and the driver. The actual stringing was supervised by a pipeline official to make sure that the pipe was closely aligned so that the lay crews would have a minimum of effort in moving the pipe into its final position. That practice later became more specialized as stringing contractors developed who bid on stringing pipe for designated sections of pipelines at prices per mile. The prices were determined by the size of the pipe, difficulty of terrain, and distance of the haul from storage to the work site.[31]

Normally, as soon as the survey was completed, the right-of way cleared, and pipe was strung, work began on laying the line. Laying large threaded pipe was an intricate and fast-moving process. It utilized two tong crews who leapfrogged each other and a score or more of pipe carriers who placed the strung pipe in the precise position for the tightening or tong crew to do its job. First, a man checked each twenty-foot joint of pipe, removed the thread protector, cleared the pipe of foreign matter, and oiled the threads. Then the carriers picked up the pipe using either "calipers" or "carrying irons" and put it in position for tightening.

At that point the "stabber" started the threads and four men, two on either side of the pipe, pulled in an alternating rhythm on two large ropes looped two or three times around the pipe in order to "spin it in," or make the connection snug. Then the tong men, using two sets of Klein tongs, generally referred to as "lay tongs," "pipe scissors," or "hooks," which weighed one hundred pounds or more and were operated by three or four men to a tong, tightened the pipe in a smooth, well-coordinated motion with one set of tongs going down to tighten while the other was going up to get another bite on the steel pipe joint. All that operated to the rhythm of the "collar pounder," "tapper," or "pecker," depending upon the vocabulary of the crew, who used a hammer to tap a rhythm on the collar or thick part of the pipe where the pipe joins. The collar pounder actually served two purposes. His tapping kept the threads from sticking by dislodging any dirt or rust embedded in them and also set the pace at which the tongs operated. As the pipe became tighter more sets of tongs were added to the process, up to a maximum of six, or three crews of men. The average laying crew consisted of approximately one hundred men who labored from dawn until dusk to get in their ten- or twelve-hour day. Because their work was considered more skilled, pipe layers earned from twenty-five to thirty cents per hour. All the work was done by hand and the average crew could lay about 175 joints of the twenty-foot pipe, or about three-quarters of a mile, per day.[32]

Once the pipeline was laid and set on skids, usually four-inch by four-inch timbers, to keep the completed line out of the dirt the ditching crew began its

*Pipeline laying crew, ca. 1914, complete with all their specialized tools, illustrating the large number of workers necessary to lay threaded pipe. (Photo courtesy of the Houston Metropolitan Research Center)*

work. Digging the ditch required only unskilled labor and the crews usually numbered around one hundred men. The depth of their excavation varied, but it was usually a ditch approximately one foot wide and eighteen inches deep. Ditch depth was often determined by burying the line beneath the depth at which most farmers plowed their fields. In some cases landowners were even contracted to do the ditching across their property. This device served as an additional incentive to gain access across private property as cash money was hard to come by in rural areas in those days. The ditching work was another of those few occupations associated with the oil business where African Americans found employment. Wages for the ditching process averaged fifteen to twenty cents per hour and a days work ranged between ten to twelve hours depending upon the combination of weather conditions and what time the sun rose and set. The laborers worked seven days a week in an era where overtime pay was an unheard of luxury.

Once the ditch was completed the pipe was raised off its wooden supports and lowered into the excavation. This was usually done early in the morning when the pipe was cool and had contracted to its shortest length. Even then the pipe was laid in the trench in as much of a side-to-side variation as possible in order to allow for heat changes that would stretch the steel and possibly cause the metal to break. The work of lowering the pipe was accomplished with either wooden hand windlasses or, on rare occasion, by the use of draft animals. Once the pipe was safely in the ground the backfill crew began its work using either Fresno scrapers or "marmon boards," a board about five feet long with ropes attached at each end so that horses could pull them to drag the dirt into the ditch. Following that the backfill crew smoothed and cleaned up the right-of-way.[33]

All of the crews involved in laying, ditching, and backfilling pipelines were housed in separate mobile camps depending upon where they were located in the process. Those crews using African American labor were housed separately. The camps were normally set up from three to five miles ahead of the lay crew so that the pipeliners walked from camp to the work location each day. As the work progressed to a few miles past the camp then the camp was once again moved ahead of the work. Thus, the camps tended to remain in the same spot for a week or ten days depending upon how fast the line was being laid. When set up and operating, the typical pipeline camp consisted of about eight or ten tents capable of sleeping ten men each, a large mess tent, a cook tent, and some sort of bathing apparatus that was usually a barrel set on a platform to be used as a shower. All that equipment was transported by four or five wagons loaded with tents, cots, cooking utensils, and food supplies.[34]

Payday came every two weeks on a pipeline job and it was always handled as cash payment on the job site. On a big job like the Glenn Pool to Humble project where the entire operation lasted six months, a central construction office was established about midway along the line. In that instance Dallas served as the location. From that office several different sets of crews operating at various points along the right-of-way could easily be served. Every two weeks the paymaster left Dallas with a valise containing approximately $20,000. He visited each of the camps and paid the men in cash. Because so many of the workers were illiterate and signed with their mark, the paying process took considerable time due to having two witnesses for each of the mark signatures. With money in their pockets, the hands immediately began to look for some sort of recreation, but, unlike being in the boomtowns, there was no place readily available to carouse. Besides, as tired as most of them were, they had little energy for that sort of activity. Card playing at night seemed to be their primary recreational activity. It seemed that in every camp a couple of professional gamblers would always manage to get hired and within a day or two after payday they usually had all the loose money in the camp gathered up and disappeared. The process on the larger jobs, as described by Mr. Hull, was duplicated on the smaller ones, with the main difference being that the work lasted a shorter time. Thus, pipeliners tended to jump from job to job and location to location as word of another big project passed among their fraternity.[35]

Payday represented another problem for the pipelining industry. There was a heavy turnover in the tong crews that usually manifested itself on payday. Many of those workers only stayed on the job for a few weeks, which was long enough to get enough money together to go on a spree. Every payday saw the loss of a number of the old cats who left to go on a big drunk. Sometimes they would return to that job and be reinstated or they would simply drift on to another job somewhere else because work was usually plentiful.[36]

So those were the oilfield hands during the early days of the twentieth century. The drillers and the roughnecks, the rig builders and the tankies, the casing crews and the pipeliners, and all the others mentioned and unmentioned here that made the oil patch work. They were then and still remain, by and large, an unsophisticated lot who drifted into the work because it paid better than farming or ranching or clerking or some other sort of humdrum traditional type of work. They stayed with the labor because it paid well, was not particularly harder than what they were used to, and because, at least at first, it was the epitome of the great adventure that every boy dreams of living some day. For those who stayed with it, some became wealthy and socialized down at the petroleum club, but most of them ended their careers in the patch much like they began it, as workingmen in greasy overalls.

CHAPTER 4

# Moving on up North, 1910–1922

By 1910 the Gulf Coast hands began to fan out across the Texas landscape in answer to the siren call of oil riches. Pennsylvania and West Virginia veterans of the oil patch along with the newly trained Texas boys began to appear in widely separated parts of the state. And everywhere they went they repeated the process of hiring local "boll weevils" to fill the incessant demand for labor. By around 1920 or so the state's petroleum industry had assumed such a dimension that Texans were acknowledged as oilmen of the first order.

Despite all that frenzied activity and its associated success much of the state's early development was accomplished by amateur entrepreneurs working with clumsy and inefficient equipment. An assortment of unlikely candidates, all of them with a get-rich-quick gleam in their eyes, entered the fray. Small-town businessmen, farmers, ranchers, and every other conceivable type of person got in on the action. Some succeeded beyond their wildest dreams, but most failed. Nevertheless, those who succeeded so spectacularly established the enduring legacy of Texas as a place where oil wells flow and millionaires grow.

One of the first discoveries that followed the Gulf Coast activity happened on the opposite side of the state. A petroleum find appeared on the Red River north of Henrietta as early as 1902 when a farmer drilling for water discovered oil. By 1907 that field around Petrolia consisted of 169 shallow wells ranging from 150 to 300 feet in depth. That year also marked the drilling of a gas well that, by 1913, was supplying gas service to twenty-three towns and cities including Fort Worth and Dallas. Although that little North Texas boom did not nearly equal the Gulf Coast frenzy, it did spawn the town of Petrolia in its midst and call attention to the possibility of further discoveries in the region.[1]

Local excitement prompted numerous attempts at finding oil along the Red River. One drilling contractor from South Texas who drilled numerous wildcats in the region was Claude L. Witherspoon. He moved a rig from the Gulf Coast up to the Denison area at Preston Mill to drill in the Red River Valley. When he spudded in on the first well about three o'clock in the afternoon the event attracted an audience of well over one thousand local citizens. Witherspoon had agreed to drill three 750-foot wells for the exceptionally high price of three dollars per foot. The local developers agreed to pay him on the basis of every 250 feet he drilled. By 10 P.M. on the day he started drilling the first well, the contractor had the well down to 500 feet and by daylight he completed it as a dry hole. To say that the promoters were dismayed would probably be an understatement. To compound the situation the wily well driller had insisted that he could drill only in the level river valley country where moving his rig from location to location would be much easier. By the time he finished his second well his employers were adamant that they renegotiate the drilling contract. By the time the entire fiasco ended Witherspoon managed to drill five dry holes at one-half the originally agreed price. He pocketed a tidy sum while the inexperienced oil promoters came up empty handed. This was destined to be an often-repeated scenario between locals and oilmen over the years.

Witherspoon followed up his Red River project by transferring to Mineola in Kaufman County. A self-styled oil finder there had convinced the town populace that he could locate oil by using some sort of a divining rod. Local citizens got behind the project. They even began to hold prayer meetings on a regular basis in order to give them a divine edge on their oil well endeavor. The contractor took the job for a one-half interest in the well plus a fifty-cent per foot drilling charge, which at the time was pretty much the going rate. Witherspoon spudded in atop a high hill just outside town; because the local oil finder had heard that oil had been found on Spindletop Hill (a place he had never seen), he reasoned that drilling should take place on a similar geographic location. Everybody in Mineola and the surrounding countryside bought a little stock in the Bluebonnet Oil Company, which promoted the well, and one Sunday afternoon after church more than 3,600 true believers turned out in person to see their prayers at work by observing the drilling operation. The local ladies even made a huge blue-colored bonnet to fit over the top of the derrick to commemorate the name of their very own oil company. But, alas, that well, like most wildcats of the day, came in a duster.[2]

There is no way of even estimating how many of those types of wells were drilled in out-of-the-way Texas places by unknown drilling contractors for trusting small town and rural folk who banded together to bring oil riches to their communities. There is very little solid information on much of the bona fide oil company activity where no discoveries were made much less the adventures of the less reputable concerns. It seems dry holes got very little press. However, judging from the anecdotal stories told by residents of counties scattered across the state in the 1900 to 1920 era, there must have been hundreds if not thousands of such unsuccessful attempts. Those that were successful were usually well documented and some of the bizarre happenings surrounding those successes give us a glimpse of the nature of well drilling across Texas during that time.

Perhaps one of the most publicized discoveries dealt with the find a few miles west of Wichita Falls at Electra. It began in 1903 and 1904 when legendary cattleman W. T. Waggoner decided to drill some water wells on his gigantic DDD ranch to offset the effects of a prolonged drought that was severely curtailing cattle production. Of the first four wells he drilled all produced only salt water laced with traces of oil. Abandoning the effort in disgust, he was quoted in later years as saying, "I wanted water and they got me oil. I tell you I was mad, mad clean through. We needed water to drink. I said, 'damn the oil, I want water.'" It was easy to say those kinds of things after the fact when oil had made him one of the wealthiest individuals in the state.[3]

Interestingly enough the showing of oil found in those 1,500- to 1,700-foot efforts created little area interest. Local folk rigged up buckets and cans tied to baling wire and drew oil, which rose to a few feet beneath the surface, with which they doped their cattle for ticks and their hen houses for mites. But, as word seeped out, oilmen quietly began leasing local property. By 1909 drilling was taking place in the area and by 1911 they had a small producer.

On April Fools Day of 1911 the Electra boom began when the Clayco #1, located about one and one-half miles north of town, came in. On the night of the discovery one of the crew telephoned Clayco's field manager to report that the #1 well was bubbling oil. Awakened from a sound sleep, the manager instructed the crew member to shut the well in and not to bother him again until after daylight. By the time the sun rose the next morning the well was blowing oil one hundred feet into the air. It covered all the surrounding vegetation with a thick greasy coating of oil. Word spread quickly, oilmen flocked to the site, Electra assumed boomtown status, and by the end of the year more than one hundred wells were in full production.[4]

Among those who made the Electra boom, the first to rival the gigantic Gulf Coast finds some ten years earlier, was A. R. Dillard. He was an experienced Corsicana and Gulf Coast rig builder and driller who was on his way to find work in some newly opening oilfields in California. On a whim he decided to stop and look over the Electra action that was being played up in all the state's major newspapers. Dillard bought a train ticket to Wichita Falls, which was rapidly becoming the regional administrative center for oil activities. Once on location it did not take long to discover that the local hangout for the oilmen was located in a sporting goods establishment conveniently situated in the same block as seven saloons. Within an hour or so of his arrival in town he gravitated there in order to get a line on the oil activity everybody was talking about.

Dillard discovered that the best way to get to the scene of the action was to take a little short-line train dubbed "Coal Oil Johnny" from Wichita Falls to Electra. Leaving every morning at seven and returning each evening at six or six thirty, Coal Oil Johnny transported most of the supplies and workers to the new field. The Electra that Dillard saw early that frosty March 15 morning in 1912 was a totally unimpressive hamlet that consisted of a few frame buildings facing the railroad tracks supplemented by several more similar structures scattered across the prairie. However, within an hour of arriving, he ran into Frank Winsett, with whom he had worked beside in a rig-building crew back in Corsicana. His old working buddy immediately began trying to convince Dillard to go to work for him as a rig builder. It seems that just a few days earlier a terrific storm had blown through the area and toppled most of the derricks in the field. The rig-building contractor was desperately trying to recruit experienced rig builders from across the state. Dillard protested that it had been several years since he had done that sort of work and that he was too soft to stand up to such hard labor. Nevertheless, Winsett persuaded him to help by offering premium wages. For about a month Dillard built rigs in the Electra field. By that time the boom was escalating into a mix of both rotary and cable tool drilling activity, and the demand for experienced rotary drillers was so great that he changed jobs in order to work for a drilling contractor named George Orr. But, as he explained it, "I failed to collect all my wages. He was kind of light on finances and slow to pay." By that time Dillard began to like that part of the country so much that he stayed around the Wichita Falls area for a number of years running rotary rigs for variety of drilling contractors.[5]

The rotary rigs that Dillard and others operated during the north and north central Texas area during the 1911–20 era remained very crude pieces of machinery, although they were much more powerful than the rigs of the Spindletop boom. By the same token the cable tool operation was little changed from the 1900 period when they reached the peak of efficiency that they were ever to attain. Both those types of drilling rigs proved very good at drilling the less than 2,000 foot wells in the Wichita Falls area and later at Ranger, Desdemona, Breckenridge, and other locations where the drilling depth remained less than 4,000 feet.[6]

One of the less well known changes to those drilling operations was the use of portable electric light plants introduced in the mid-teens by companies like Delco-Light and the Lucey Manufacturing Company. Those gasoline-powered light plants replaced the old yellow dog lamps, kerosene lanterns, and the steam-powered generators of the early oilfield and eventually became the mainstay of drilling rig lighting.[7] Despite the obvious disadvantages of open flame lighting around drilling rigs some refused to change. Carl Mirus commented on the situation.

> For a long time lots of the drilling contractors wouldn't put electric lights on their drilling rigs. They used to use these old "yellow dogs." It's a cast-iron teapot with a double spout. If you fill it up with crude oil and stuck a piece of toe sack in each spout and lit it, it was a double torch. That was the lighting system on a drilling rig at night. Three or four of those, one

> over the driller, one over the engine. They used them, oh, I'd say some of the more conservative drillers used them up until about 1925. If they struck gas they were goners, but that was just a hazard of the business.[8]

A. R. Dillard and the hundreds of oilfield workers like him who descended upon the Electra boom found plenty of work as minor discoveries continued to be made in the area. Later, the big Burkburnett find just northwest of Wichita Falls and then the less spectacular field at Iowa Park cemented the region as a major player in the Texas oil story. Wichita Falls, located midway between Petrolia and Electra, blossomed from an obscure railroad town into a fast-growing hub of oil activity in northern Texas. Despite the town's booming aspect no oil was found in the immediate vicinity of the municipality. To solve that problem the town fathers posted a $10,000 reward for the first producing well drilled within six miles of the town site. But they were doomed to disappointment for years to come.[9]

However, less than twenty miles away on July 29, 1918, a wildcat well on the S. L. Fowler farm, which joined the town site of Burkburnett on the north, blew in at 2,200 barrels per day. The boom was on! Within three weeks more than fifty rigs were drilling in the immediate vicinity of the discovery well. Thousands crowded into the little village of Burkburnett, and nearby Wichita Falls became the center of a speculative mania centered on the sale of oil stocks both legitimate and less than above board.[10]

The Burkburnett boom, combined with a similar situation at Ranger beginning six months earlier, created a tremendous oilfield labor shortage in the region. The situation was exacerbated by the effect of the World War I military draft that had drawn significant numbers of trained oilfield workers into military service. This reached a critical stage in the fall of 1918 when as many as forty rigs in the area were idled due to a lack of trained personnel to operate them. Although the government authorized furloughs from the army for qualified workers to return to their oilfield work, very few took advantage of the opportunity. Consequently, there was another surge of very young men with a rural background entering the labor force.[11]

The discovery at Burkburnett developed when a discouraged farmer named S. L. Fowler decided to leave his unproductive acreage and move on to greener pastures. He was dissuaded from giving up altogether by his wife; she argued that because they were ruined anyway they might as well try to interest someone in drilling for oil on their property in a last ditch effort to salvage something from their unfortunate situation. Fowler managed to scrape up enough funding from friends, family, and a couple of Gulf Coast oilmen to begin the well. Walter Cline was one of those experienced oilmen who got in on the Fowler well when he contributed his rotary drilling rig and a crew to operate it for an interest in the proposition. Cline, who had recently moved his drilling contracting operation from the Gulf Coast to Wichita Falls to take advantage of the growing opportunity there, made the single most important deal of his life when he invested in "Fowler's folly," as the well was known locally.[12]

Once again, following the pattern of development initiated during the Gulf Coast boom, chaotic conditions reigned supreme in the field around Burkburnett. Wichita Falls emerged as the more stable administrative center where a massive speculative stock business thrived. It was very similar to the relationship between Beaumont and the more rowdy situations at Spindletop, Sour Lake, and Batson. Thornton Lomax observed all this when he made the Burkburnett boom as an oil scout and lease peddler. He normally went out into the field at five or six o'clock in the evening and lay concealed in the brush on some hillside using binoculars to spy on rigs in order to ascertain how their work was progressing. By counting the joints of pipe as they were pulled out of the hole he managed to estimate the drilling depth of particular wells. If the project happened to be successful Lomax immediately attempted to lease as much of the surrounding land as possible. In this manner he would sell those leases in a day or so at enormous profit as news of that particular find leaked out. It was not unusual to secure leases for as little as two dollars per acre and sell them for as much as thirty-five dollars. On one memorable occasion Lomax sold $50,000 worth of leases on the streets of Wichita Falls in a single day.[13]

Hundreds of speculators, similar in nature to Lomax, operated in Wichita Falls. Business was so brisk that the county clerk added twenty or more clerks to his courthouse payroll in order to handle the massive number of land transactions. There were dozens of stock exchanges in town that dealt exclusively in oil stocks; there were also hundreds of "lease hounds." But not all of those operations were as ethical as they could be. Landon Haynes Cullum described some of the activities ascribed to those shady characters.

> They put out some of the darndest maps you ever saw in your life. Why, those maps had them owning producing wells right in the middle of the field while the stuff they were selling wasn't within ten miles of any production, and they were making all kinds of wild promises and statements, and it was awful.[14]

One of the promoters' favorite schemes was to get a lease on land that had no possibility of oil production but which lay a considerable distance directly between two known producing areas. The speculator would laud his leases as sure producers that were surrounded on all sides by proven oil production. Another popular scheme was to sell several hundred percent interest in a wildcat prospect so far from known production that it was certain to be unsuccessful. That particular activity was known to come to grief when the overcapitalized well was

*Street scene in Burkburnett, ca. 1919, illustrating the mix of horse-drawn vehicles and newly introduced automobiles operating in the unusually wet weather of that year. (Photo courtesy The Petroleum Museum, Midland, Texas, James Flowers Collection)*

accidently successful and the investors arrived to collect their money from the promoter, who usually left the country in a cloud of dust. A more common practice was to overprice the stock in a small operation, usually a one well deal, in order to pay for inflated drilling expenses so that the speculator would make his money on the actual drilling of the well, and the investor would never get a return on his investment. Then there were the outright frauds who sold stock in nonexistent leases and companies. They were all present in the Burkburnett boom and they all did a lively business there and on the streets of Wichita Falls.[15]

Promoters and speculators aside, there were a tremendous number of oilfield hands laboring in the area. The Burkburnett boom was primarily a rotary rig drilling boom that attracted large numbers of the Corsicana and Gulf Coast boys who were experienced in the rotary drilling trade as well as some of the old-time cable tool men. Depending upon how far the rig was from town the drilling crews either stayed in shacks at the drilling site or had rooms in town. Alexander Patterson, who worked on some of the cable tool wildcats in the area, remembered his quarters as consisting of a one room shack with a double bed. The night driller and his tool dresser shared the bed. When they went on tour the daylight driller and his toolie slept in the same bed. The room was also shared with the cook who slept on a cot. Later, most companies provided separate quarters for the various shifts working at such isolated locations.[16]

As greater numbers of men began working for the larger firms the oil companies built bunkhouses and dining halls similar to the situation at Spindletop. Additionally, it was a time when those companies began to build clusters of permanent housing, usually called camps, in order to attract and keep more settled and reliable family men on their payroll. A report on a study made by a committee of the Texas, Gulf Coast, and Louisiana Oil and Gas Association made during this period indicated that it was imperative that the large companies implement such programs for the stability of the industry. Examples like the Gilliland Company cantonment at Burkburnett, the Prairie Oil and Gas Company's camp at Ranger, and the Humbletown camp at Cisco soon developed across the area.[17]

However, the Burkburnett boom was dominated by a myriad of small operators who could not afford to build such luxurious accommodations. Consequently, most workers in the area were forced to sleep in flimsy rooming houses or in large tents crowded with cots where for $3.50 you had the undivided use of a cot for twelve hours. Those establishments then rented the cots to the next shift without changing the linens or cleaning the place in any other way.

All other businesses catering to the oilfield workers operated under similar constraints. It was next to impossible to get laundry done in the midst of the boom. Consequently, most of the hands bought new work clothes, wore them until they were so filthy they could not stand them, and then discarded them and bought new apparel. Meals also presented problems just as they had on the Gulf Coast. Boardinghouses seemed to have provided most meals, although it was difficult to get inside and find a seat. The alternative was to eat at one of the local cafés where the meals were usually less than appetizing. You usually paid at the door when you entered one of those establishments whether you intended to eat or not. Patrons entered by one door and exited by another. Entrances and exits were guarded by armed men who assured that dining rules were strictly enforced. When the patron's food arrived it may or may not be what he had ordered. The best procedure was to eat what was set before you; if one protested that something was wrong with his meal the plate was given to someone else and the unfortunate soul who lost his dinner was summarily escorted from the establishment by one of the guards.[18]

When Jack Knight arrived on the scene in 1918 he frequented the Mecca Café in Wichita Falls. There you got a glass of water with your meal, but if you did not tip the waiter fifty cents you got no more water. Actually, potable water was at a premium during the boom. Many workers would tip the waiters as much as a couple of dollars simply to get as much water as they wanted. To get a bath cost from $1.00 to $1.50 for fifteen minutes in a bathhouse, after which the hapless patron was unceremoniously ejected in favor of another eagerly waiting customer. The large number of bathhouses did a thriving business from a clientele that spent long days working in a very dirty environment.[19]

The summer of 1918 was hot, dry, and dusty along the Red River in northern

Texas, but that fall it began to rain and mud became a real hindrance to oilfield activity. Although not nearly the magnitude of problems that developed in the swampy Gulf Coast region, wet weather at Burkburnett created hazards that definitely slowed the pace of activity. The massive amount of traffic through the town made Main Street so impassable that two blocks had to be cordoned off to all traffic. Every stream and low place in the oilfield became a potential place for vehicles to bog down. Consequently, toll bridges cropped up everywhere there was a likelihood for mud holes. One company found it such a bother that they overcame the problem by allowing contractors to build bridges across all low places and charge twenty-five to fifty cents per vehicle for all those using the bridges. Company employees who crossed free of charge were the exception. Prices varied of course, but fifty cents per person and $2.50 per truck were common charges for toll bridges. The practice developed to the point that between Burkburnett and New Town, a distance of seven miles, there were seventeen toll bridges operating at one time. The alternative to paying the toll was getting stuck in the boggy places. Depending upon the situation it could cost from five to fifty dollars to get pulled out of those places.[20]

At Burkburnett, although motorized trucks were coming into vogue for hauling freight, teams of horses and mules were by far in the majority within the confines of the field. No ox teams were used in that area, as was the case in the earlier Sour Lake/Batson situation on the Gulf Coast. Motorized trucks are credited with moving 95 percent of the freight from the railroad yards at Wichita Falls to Burkburnett, where it was usually transferred to wagons.[21] When the trucks got stuck, as was often the case, the teamsters pulled them out for whatever fee they could extort. That activity led to a considerable rivalry between the truckers and the "long line skinners," as the teamsters became known. It was not unusual for a truck and a freight wagon to meet on a good stretch of road and for both refuse to give way and allow the other free passage.[22]

Witnesses claim that those standoffs could go on for extended periods of time and many times resulted in fistfights.

> They had a few trucks in there, and those trucks would take down the road, and there would come a bunch of teams moving some equipment and when they met, well, right there is where the clash was. Those teamsters wouldn't pull out for them truck drivers and those truck drivers couldn't pull out and they'd just set there. I've seen them set there all day waiting for—one of them waiting for the other one to pull around him. Maybe they would get into a doggone fight and just fight it out, and then one of them would give up to the other one and they would move around. . . . They'd try to maneuver by each other, those truck driv-

> ers would. They'd get along fairly good with each other, but when they hooked up with those long line skinners that's when they'd fight.[23]

As usual in the oil patch wages were good in 1918. Roughnecks and other skilled hands averaged ten dollars per day and unskilled labor was rewarded with above average wages according to the type of work being done. As in preceding booms this atmosphere created a situation where a host of workers were parading around looking for a good time and flaunting their money. Naturally, this attracted a considerable lawless element who saw this activity as a golden opportunity. The most publicized criminal activity during the Burkburnett boom was armed robbery, or *hijacking*, as it was termed then. Most of the hijacking happened at night in alleyways and on dark deserted streets in simple face-to-face confrontations, although a favorite scheme was for two or more criminals to go out in the middle of the night to where a rig was running and rob the entire crew while they were on the job.

Some of the more enterprising stick-up artists made a practice of preying on groups of men just after payday. They would enter the tent where the men were sleeping and herd them to one end of the tent. While one gunman stood guard over the sleepy victims another would ransack their possessions and take any valuables. One particular group of hands who had been robbed on three separate occasions became much more careful and managed to capture their tormentor. Upon investigation the thief proved to be female. Not being prone to doing harm to a woman, the workers decided to strip her naked and throw her out into the muddy Main Street of Burkburnett for all to see.

Local law enforcement officials who were temporarily overwhelmed with the volume of criminal activity that seemed to have sprung up overnight seemed unable to control events. Consequently, workers many times took the lead in solving the problem in their own direct way. One incident that illustrates this was the case of a company guard who decided that enough was enough and proceeded to stop the hijacking activities that were plaguing his section of the field. The man bandaged his hand and arm in a fake splint where he hid a pistol. Then he paraded from bar to bar making a great show of having a large roll of bills. As he returned to his rooming house late that night a hijacker accosted the supposed victim. That would-be thief was killed for his trouble and the problem abated for a while.

In addition to armed robbery there was a tremendous amount of theft of oilfield equipment. The problem escalated to the point that most oil companies hired armed guards to watch over any equipment not stored in a locked warehouse. The thieves became so bold that on one occasion an engine was stolen off an idle rig at one end of a small lease while at the other end of the property the watchman filled out his paperwork before making his next inspection round. Thieves would steal drill pipe right off the pipe rack adjacent to an operating rig at night and quickly put enough distance between them and the scene of the crime so that their tire tracks would become lost in the jumble of other similar tracks in the main traveled road when the owners followed their trail the next morning. Perhaps the safest form of theft, which was practiced by large numbers of otherwise honest hands, was to throw expensive new drilling tools into the rig's slush pit and tell the boss that they had disappeared. After the well was completed and the pit dried up the culprit would return, remove the buried tool from the dried mud, and sell it.[24]

While all those nefarious activities were happening up on the Red River around Wichita Falls an equally spectacular situation had developed about a hundred miles to the southwest. Exploratory activity in that region began as early as 1890 when Abilene residents unsuccessfully tried to find oil. Then the influence of the Spindletop discovery prompted another and more extensive round of inquiries in the central West Texas area, but it was not until 1912 that any of the efforts met with success. In that year the Texas Pacific Coal Company at Thurber found slight showings of oil while drilling test holes for coal. Encouraged by the positive signs, the coal company began to drill for oil and three years later they completed a successful well at Strawn. Then in 1916 Texas Pacific drilled several more small producers in both the Ranger and Breckenridge areas. Early the following year the company was so encouraged by its growing success in the oil business that the firm changed its name to the Texas Pacific Coal and Oil Company and, at the insistence of the good citizens of Ranger in Eastland County, leased 25,000 acres of land in the immediate vicinity of that unincorporated village.

That spring Texas Pacific Coal and Oil spudded in on Mrs. Nannie Walker's farm located on the north edge of Ranger. At 3,400 feet they got a strong flow of gas. However, before they could complete the well they lost a string of tools in the hole and were unable to retrieve them and had to abandon the effort. This forced them to begin another well, this time on the J. H. McKlesky place about two miles south of town. W. K. Gordon, general manager for Texas Pacific Coal and Oil, had developed an intense case of oil fever by this time. Although ordered by his superiors in New York to discontinue drilling if oil was not found at a maximum depth of 3,200 feet, he ignored the directive. His persistence was rewarded when on October 21, 1917, at a depth of 3,431 feet, the McKlesky #1 blew in at a rate of 1,700 barrels per day. Once again the boom was on![25]

The McClesky well was drilled by a cable tool contractor who used drillers Frank Champion on days and Jake Walters on nights. On the day the well came in Champion was watching his drilling closely and checked the cuttings every few minutes as he neared the 3,400 foot level. At 3,431 feet he smelled gas and tasted the cuttings, which were saturated with oil. The driller pulled his drilling string up the hole about a hundred feet. Oil began to gush from the hole until it eventually rose high enough to strike the temper screw suspended about ten feet above the rig floor. The driller quickly extinguished the boilers to guard against igniting the well and walked the two miles into tiny little Ranger to report the completion of a successful well. Pandemonium broke out upon his announcement as a group of jubilant businessmen piled him into a car and drove the excited driller back to the well site. They were closely followed by an entourage of almost all of the town's eight hundred residents who were eager to see what their efforts had wrought.

What those efforts had wrought was to yank a sleepy little cotton-growing and ranching community from peaceful obscurity into the national limelight as a place of instant wealth. That image was never more striking than out at the McClesky place on the day of the discovery when oil spouted over the derrick's crown block and began drenching the adjacent peach orchard and covering all the chickens. Meanwhile, Mrs. McClesky worked frantically to get her washing off the clothesline before it was ruined by the blowing cloud of oil as throngs of townspeople milled around speculating on the extent of their newfound wealth. That scene, probably more than any other, illustrates the existing lifestyle of the area and gives a portent of the changes soon to come.[26]

Like all the other boomtowns before it, Ranger quickly began bursting at the seams as thousands rushed into town. Within weeks it went from a sleepy little country village of fewer than a thousand to 10,000 or 12,000, and within the year more than 40,000 citizens received their mail at Ranger. As usual there were hardly any sleeping accommodations, prices were high, and food and other necessities of life were scarce.[27] Carl Mirus, who arrived in Ranger a week or so after the discovery well came in, was particularly impressed with the poor food and contaminated water. He stated that "there was a saying up there that the only two safe things to eat were hard boiled eggs and coconuts, provided you had your own hammer to crack the coconut and we bought #4 Crazy Water by the case and carried it in our car to drink. We were afraid to drink the water up there, though most people did drink it."[28]

A. J. Thaman remembered that potable water was extremely hard to acquire despite the unusually wet season experienced by the region that year. He stated:

> And some of the other hardships at Ranger was the water supply. I remember that we bathed very infrequently because of the inadequate water supply. Saturday was usually bath day, and we got our water by—from an old fellow who would come by the store. Each Saturday morning he would drive up there with a team and one of those buckboard or farm wagons with eight or ten barrels of water in it, and we would go out and pay fifty cents for about five gallons of water, enough that we could put

> into a galvanized washtub. And that's the way we got our Saturday evening bath. Our drinking water was shipped from Mineral Wells. And we'd go uptown and buy a soft drink of course for the regular price of five cents, and ice water was also five cents a glass.[29]

Even the *Oil Weekly,* usually quiet on such matters, commented on the unusually difficult situation concerning the lack of water at Ranger.

> It is reported that those who have been in Ranger since the oil development started have practically forgotten what it is to have a bath, and some of them make monthly visits to Fort Worth and Dallas for the sole purpose of moistening their joints in a porcelain tub. For several months last year there was not water enough in Ranger to take a bath, and those who dared mention such a luxury were told that Mineral Wells was the nearest bathing resort.[30]

Those who made the Ranger boom maintained vivid memories of the muddy streets. Actually, the area had been going through a drought with a much smaller than average annual rainfall, but, like at Burkburnett, the winter of 1918–19 arrived exceptionally wet and cool. The unusual amount of moisture combined with an overwhelming quantity of oilfield traffic immediately created a massive problem for the state's newest boomtown. All the traffic that funneled down the

*Main Street in Ranger, ca. 1918, where throngs of teamsters crowded the narrow thoroughfare in typical boomtown fashion. (Photo courtesy Southwest Collection/Special Collections Library/Texas Tech University, Jack Nolan Collection)*

long Main Street created so much mud that it was almost impossible to navigate the street crossings. Even during the short intervals of good weather very little cross traffic could get across the main thoroughfare. In wet weather it proved impossible. Ranger received a certain amount of notoriety as the place where, for ten to twenty-five cents you could get ferried across Main Street on a mud sled operated by some of the enterprising local folk.[31]

Oil well drilling in the Ranger area was almost exclusively of the cable tool variety. Consequently, very few of the Gulf Coast boys made the Ranger boom. With their penchant for rotary drilling they tended to stay on the coast and in the Red River Valley where their expertise was better appreciated. The Ranger drilling hands coalesced around a core of old West Virginia cable tool drillers, with a few from various other places. Much like other oil boom situations local farm and ranch boys flocked to the action, eager to earn the good wages available in the oil patch. This time, however, a significant new factor entered into the employment picture. As the boom peaked in 1919 the labor shortage abated somewhat as large numbers of World War I veterans were released from military service. Those young men, restless and trying to adjust after the excitement and constant change experienced during the war, found that working in the new world of the oil patch well suited their changed attitudes. Thousands of those young veterans changed their military adventure into an oilfield adventure during the 1918–20 period.[32]

As at other booms before it Ranger had more than its share of lawlessness due to overcrowding and an abundance of easy money. Bootlegging, hijacking, prostitution, and gambling made their appearance as the undesirable element that always followed when a boom arrived. The hijacking, which became a particular problem for the hands on their way to and from the rigs, was described by one driller who bought a pistol for protection. After being hijacked the man told his friends that "it was a good thing I didn't have my gun along. They would have got that too."[33] The major criminal activities in Ranger came to an abrupt halt in 1922 when the Texas Rangers raided a notorious gambling operation at the Commercial Hotel, which prompted the local Rotary Club to establish a good government movement that successfully spearheaded the efforts that drove the criminal element out of town. Actually, criminal activity during all of the oil booms, although very real, got much more play in the newspapers than was particularly warranted because the lurid has always sold more papers than the mundane. Such journalists as Boyce House, who served as a reporter on the Ranger newspaper and wrote the book *Were You in Ranger,* along with other authors who wrote widely read books that played up the spectacular lawless aspects of life in the boomtowns, created a vivid and enduring modern-day Sodom and Gomorrah image in the public mind.[34]

Despite its gigantic oil production anchored by the large core leases of Texas

Pacific Coal and Oil and augmented by the frantic efforts of independent operators on the fringes of the field, the Ranger boom did not last long. Overdrilling and depletion of field gas pressure quickly caused most wells to cease flowing, which forced them to go on the pump at a lower production rate. Dry holes soon appeared as the limits of the field began to be probed. The town of Ranger voted for gigantic civic improvement bonds with the expectation of continued oil production. But by 1922 the boom was over, the town was deserted by most of the boomers, and city fathers were left with the heavy burden of a bonded indebtedness that the tax base could not sustain.[35]

The flurry of leasing activity at Ranger during the spring and summer of 1917 attracted thousands of lease promoters. They worked the local area extensively and soon began to fan out across the rest of the region. Many of them achieved significant success, as in the example of Landon Cullum who promoted the spectacular find at Desdemona. His adventures in that particularly chancy situation serve as an excellent example of the ingenuity and persistence of the lease hounds who worked in the oil patch in those heady days.

Cullum and petroleum geologist Bill Wrather quit their jobs with the Gulf Oil Corporation in the Wichita Falls area in early 1917 with the intent of launching out and buying leases on their own. Each had saved about $400 and they took those modest funds to a new discovery near Walters, Oklahoma, a few miles north of the Texas state line where they leased two eighty-acre tracts for two dollars per acre. Within a week they sold their two leases for eight dollars per acre and the thrill of the ease of getting such a good return on their investment hooked them on speculation in the oil patch. Immediately upon completing that initial deal rumors reached Wichita Falls of the Ranger prospects and the newly baptized lease men left for the scene of the action. By the time they arrived in town it was so full of oil promoters that "they had to wear badges to keep from selling to one another," which drove the price of the leases well beyond the modest means of the two young men. Reassessing their situation the two young entrepreneurs decided that they should travel around the countryside at a distance from the frenzied activity and try to lease wildcat acreage that geologist Wrather deemed likely to be ripe for oil production.

In the course of their investigations Cullum and Wrather drove to Stephenville, Comanche, DeLeon, and numerous other smaller rural communities south of the big Ranger boom. Although they found a couple of fair-looking sites, nothing of significance appeared. Finally, they spoke with a Dr. Snodgrass in a tiny community called Hogtown by the locals but which sported the much more grandiose name of Desdemona on the maps of the day. The good doctor put them in touch with "Shorty" Carruth, town barber and well-known local eccentric, who had persuaded local citizens to finance an unsuccessful oil finding attempt several years earlier. Shorty immediately tried to interest them in financing his second well, which he had down to two hundred feet at the time. The partners declined the offer and got away from him as fast as possible.

Dr. Snodgrass told Cullum that Shorty's first well had shown some oil sign when it was abandoned at seven hundred feet. The doctor volunteered to guide the oilmen to the abandoned well site where they recovered some oil from the hole by lowering a bottle into the well on a string. Excitedly, the lease men puzzled over how to determine local geologic structure without benefit of alidade, plane table, or other tools necessary to accomplish such work. By talking to local farmers they discovered that a layer of bluish limestone was present in all the local hand-dug water wells. The lease hounds then simply went around the countryside and let a weighted string down the water wells to determine the depth of the limestone layer beneath the surface. Using that as a rough guide combined with surface outcrops along the banks of Hog Creek, from whence the community derived its name, they discovered that a large anticline, which many times indicated the presence of oil beneath the surface, existed in the area.

Buoyed by Wrather's assurance that they had a good bet for an oil discovery, Cullum began the work of leasing drilling property. Having no funds complicated the process, but resourcefulness has always been the hallmark of good lease buyers. He called all the area farmers together at the little local schoolhouse on the night of October 1, 1917. At that meeting he informed them that they had a fair prospect of getting an oil well in their community, which would greatly benefit everyone present. Then the persuasive Cullum proceeded to lease ten or twelve farms totaling some three thousand acres for the unbelievable low sum of one dollar per farm, an investment of only ten or twelve dollars. Over the next few days he even managed to double the size of those holdings to six thousand acres.

The partners rushed back to Wichita Falls and found some backers who agreed to underwrite the cost of putting down a well. The deal stipulated that the underwriters would get 75 percent ownership in that one well. While Cullum oversaw the beginning of the drilling operation, Wrather traveled to Pittsburgh where he persuaded famous oilmen, Mike Benedum and Ross Parriott, to underwrite development of the entire field for 75 percent of three thousand acres as well as paying for the cost of drilling the first three wells. That deal left 25 percent of three thousand acres to Cullum and Wrather, which they agreed to share on an equal basis with their Wichita Falls backers. These backers were now released from the expense of drilling the first well because Benedum and company had agreed to assume that expense on the first three wells. This fast-moving and complicated transaction clearly illustrates how two determined young men managed to get a lease on three thousand acres and a share in another three thousand acres as well as making arrangements for drilling on those leases with the investment of only a few dollars.[36]

Pete Hoffman, a big bear of a man from West Virginia who was described by Cullum as looking "more like a gorilla than a man," was hired as the drilling contractor on the discovery well at Desdemona. He brought a string of cable tools to the Joe Duke farm just outside town and spudded in during early May of 1918. "One Eyed" Jack McCleary was drilling days and "Yellow" Young handled the night tour. Progress was agonizingly slow on the well, but about three in the morning of September 2, 1918, she blew in on Young's tour. Unfortunately, coals in the tool dressers forge ignited gas belching from the well. As towering flames lit up the early morning sky the rig collapsed into a heap of smoldering timbers. It took three spectacular days using the traditional steam smothering method to extinguish the burning Duke well fire and convert it into a thousand-barrel-per-day producer.[37]

That discovery well caused an instant boom at Desdemona, but this time it became a small independent producer's field instead of being under the nominal control of one company, as at Ranger. Within a year of the discovery uncontrolled drilling and the practice of venting all gas to the atmosphere began to seriously affect the production capacity of the field. Cullum and Wrather sold their part of the three-thousand-acre lease where the Duke well was drilled for two million dollars. Two years later they disposed of the rest of their Desdemona holdings for something over two million dollars.[38]

The Desdemona field's main production area enclosed an acreage about three miles by three miles. Consequently it was drilled up rapidly. The tiny community, like others before it, mushroomed into a tent and shack town serving several thousand workers. Most of the drilling crews lived in small camps established adjacent to their work sites. The wells were so close to town that it became a common practice to shut the rigs down at noon and walk into town for a hot lunch. Perhaps the most memorable social event of the yearlong Desdemona boom started from that noontime practice. One of the drillers got into a tiff with the Greek owner of a local café and the restaurateur attacked the driller and had him arrested. Whereupon the arresting officer struck the oil worker with his gun, knocked the guy's eye out of its socket, and threw him in the local jail without any medical attention. Incensed by the incident and in need of a bit of recreation an estimated one thousand oilfield hands descended upon the town, liberated their comrade from the makeshift jail by overturning it with the use of the winch of a nearby drilling rig, and began congregating at the café. The terrified Greek managed to escape relatively unscathed, but the mob destroyed his business establishment as well as an adjoining clothing establishment that had the reputation of charging exorbitant prices for its goods. Evidently the combination of bad food, bad service, and bad treatment so prevalent in all the booms boiled over into abrupt action during a brief but memorable moment at Desdemona.[39]

As in the Wichita Falls and Ranger areas, 1918 proved to be an unusually wet year. Hundreds of mule teams labored day and night to keep the rigs supplied, and, as usual, all that traffic churned up the roads and they soon became quagmires. Additionally, waste was so prevalent that on at least one occasion the road between Desdemona and DeLeon was blocked for a time by a three-foot-deep flow of oil from nearby production. It was not unusual for five or six loads of supplies to be bogged down at any given moment on the short stretch of road leading into the production area. So much gas was flared from the wells that residents had to close their window shades at night to make it dark enough inside for them to be able to sleep. If that were not bad enough the deafening roar of escaping gas combined with the constant throbbing sound of the drilling operations added to the nighttime discomfort that made getting a good night's sleep a real challenge. Given the small size of the field combined with intense drilling and profligate waste, it was not at all surprising that Desdemona returned to its status of little Hogtown by late 1920, when the bulk of the boom community moved on to greener pastures.[40]

One of those greener pastures was Breckenridge. It boomed from a population of six hundred at the time of the first big well in February of 1918 to an estimated high of 35,000 or 40,000 within the year. It was said that there were two hundred derricks working inside the town in that year and eventually more than two thousand could be seen from the top of the courthouse. But like all the others it lost its luster within three or four years and lapsed back into a more sedate lifestyle.[41]

Oil well drilling in that region was done almost exclusively by the cable tool method, so there was hardly any demand for the experienced rotary hands from the Gulf Coast like there was in Burkburnett area. All this cable tool work attracted a flood of West Virginia and Pennsylvania drillers to the area. Many of them were colorful characters whose antics have become part of oilfield lore. Grant Emory serves as a good example of the taciturn, hard-bitten, cable tool driller of the day. On one occasion while drilling nights he and his tool dresser had put in several hours of work without a word passing between them, then a thunderstorm began to blow up out of the west. The toolie said, "Well Mr. Emory, looks like we will have to go home in the rain tonight." Emory just sat there on his drilling stool and stared at his helper, and after a long pause answered, "Young man, you talk too much." He was noted for giving short and succinct answers to questions if he deigned to answer at all. On another occasion at a Ranger boardinghouse the proprietress noticed that Emory was slathering great slabs of butter on his biscuits. She decided to intimidate the driller by saying in an accusatory tone, "Mr. Emory, did you know that butter is costing me seventy-five cents a pound?" To which Emory replied, "It's worth every cent of it," and kept right on munching his food.

In his later years liquor got the best of Grant Emory, and he spent most of

*Breckenridge in 1921, complete with rigs operating within a stone's throw of downtown, where teamsters and automobiles share the road. (Photo courtesy Southwest Collection/ Special Collections Library/Texas Tech University, C. C. Rister Collection)*

his time and effort trying to get just one more drink. On one occasion two of his drinking buddies from back east, "Yellow" Young and "Toledo" Jack, made the rounds of the oilmen in Breckenridge collecting funds to bury poor old Grant, who they said had gone to his reward the previous night. One man donated twenty dollars to the cause. When he was downtown later that day he spied the money collectors along with the "deceased" staggering down the street with their arms wrapped around one another and singing at the top of their lungs in an obvious state of complete intoxication. Some two years later the same two collaborators approached the oilman once again about collecting for the same cause. This time the oilman cut his donation in half. He didn't mind donating to their entertainment but he felt they could probably get by on less.[42]

Another of those West Virginians who made all the booms in that region was Pete Hoffman, who contracted drilling the Joe Duke discovery well at Desdemona. Pete enjoyed a far-flung and well-deserved reputation for "borrowing" oilfield equipment without necessarily getting permission from the legal owners of said equipment. A story that crops up time after time in the early oilfield yarns concerns the stealing of a steam boiler with the fires lit and the steam pressure up. That particular stunt has been credited to Pete Hoffman, who supposedly hitched his team to a wheel-mounted boiler, disconnected the steam lines, and drove off with the equipment while the drilling crew was elsewhere enjoying their lunch. Pete was in the habit of taking so many bits and other relatively small pieces of equipment, and burying them until he could return at a later date to retrieve them, that it became a common practice for the local oilfield hands to refer to any freshly dug piece of ground as one of Pete's hiding places.[43]

On one occasion Hoffman was dragged into court and tried for stealing an eight-inch cable tool drill bit. The evidence showed that the thief had halted his wagon some distance from the rig and had somehow transported the heavy bit from the rig to the wagon. Pete's attorney did a fine job of convincing the jury that the average man could hardly budge the heavy bit and nobody could carry such an object the distance his client was alleged to have done. For added emphasis the bit in question was even displayed in the court room. After the jury of twelve men good and true found Hoffman innocent the big hulking oilman asked the judge if the bit was now his property. Upon being told that the bit indeed belonged to him, Pete bent over, picked up the massive steel object, and walked out of the court room.[44]

Characters like Grant Emory, "Toledo" Jack, "Yellow" Young, and Pete Hoffman, who ended their days in the Breckenridge/Ranger/Desdemona area, represented an era that was fading in the 1920s as those booms began to wind down. They were representative of the independent, cantankerous, early day cable tool drillers who cut their teeth in the early Pennsylvania/West Virginia oilfields and followed their trade to the oil booms of Texas. In the case of Hoffman, his actions, although beyond the norm, grew out of the practice of oilfield borrowing. It was customary for most deals between oilmen to be oral agreements sealed with a handshake. During the excitement of flush production the drilling hands on the rigs who needed more equipment simply borrowed from nearby rigs. Driller Ed Matteson recalled one instance when he was out on location and began rigging up only to find his steam boiler missing. Upon investigation he discovered that the crew down the road had borrowed it when theirs was damaged. Within a day or so Matteson received a new boiler out of a shipment his neighboring crew had just received on an incoming train. That sort of reciprocal use of oilfield equipment put most of the hands in a sort of shadowy half-world of legitimate ownership. This was what Pete Hoffman depended upon. He always said he was just "borrowing" the missing items found in his possession, and in a gesture of friendship always offered to return them once he was caught with the goods.[45]

The period from around 1910 into the early 1920s saw the expansion of the Texas oilfields away from the Gulf Coast and across the state. The most spectacular finds during the period were concentrated in the north central part of the state where booms developed at Electra, Burkburnett, Ranger, Desdemona, and Breckenridge as well as at numerous less well known locations. All those places developed overnight into boomtowns similar to the situation at Spindletop a few years earlier. For the most part the drilling activity, with the notable exception of Burkburnett, spurned the newer rotary rigs for the tried and true cable tool drilling technique. It was spurred on by the energy demands of World War I and the proliferation of the automobile in the United States.

During this period clearly defined characteristics of oilfield workers began to develop as perceived by the workers themselves. The hands, those who actually performed the labor, began to be represented more and more by local boys and those returning from the war, while all those imported oil experts from the older fields in the East began to play lesser roles. Although the makeup of the workforce changed, the egalitarian and independent nature of the workers remained an ingrained force in an occupation where stories abounded of workers who, with a little ingenuity, became wealthy. This attitude developed within a free and easy framework where a handshake sealed many a million-dollar deal and an atmosphere of mutual cooperation abounded.[46]

Technological change also began to play an increasing role during this period. There were fewer oilfield fires as open flames for lighting and better well flow control mechanisms were invented. Rotary drilling, away from the unconsolidated formations of the Gulf Coast, made some headway as heavier duty well-drilling equipment appeared that allowed them to drill deeper and faster although there remained a strong antirotary bias. Trucks and automobiles began to make inroads into the freighting business although horses and wagons continued to dominate the trade. In general there was a gradual change toward mechanization that made the industry more efficient and required a slightly more knowledgeable worker.

It was also during this period that the major oil companies as well as many of the larger independents began to emphasize a more scientific approach to various aspects of the oil business. More highly educated managerial personnel entered the operational end of the business as technological innovation became more important. Perhaps one of the more important innovations was the acceptance of the petroleum geology profession as a serious aspect of the industry. When wells began to be drilled beyond that 1,000- and 1,500-foot range, a knowledge of surface geology became less and less important and professionals who understood subsurface geology as a means of predicting likely places to drill became essential.

All this indicates a certain maturation process of the petroleum industry that is often clouded by the more spectacular aspects of the concept of a boom. As more and more large oil companies developed they added an element of social control. One does not make money on a long-term basis if one cannot exert a modicum of predictability and control over the elements of the income, and that is what the large companies did. They hired professionals, provided housing and stability for their workers, and attempted to gain as much control over oil production as possible. In the process those workers tired of moving from boom to boom aspired to work for those entities and inadvertently created a sort of social status within the working community. The company men were perceived as having a slightly higher social standing than the contract workers. Conversely, the contract hands looked upon the company men with a certain amount of disdain for having traded their independence for security.

The problem with identifying the various worker social divisions within an oil boom situation other than in a general way between the company men and the contractors is that the lines separating them were fuzzy and they constantly fluctuated. The contractors and their workers went from boom to boom in order to capitalize on the easy money during the short-lived excitement. Many of them tired of the excitement and dropped out of the cycle to settle in a particular location, but did not necessarily begin to work for major oil companies. By the same token many oil company employees were transferred from location to location as new discoveries developed. There were also a host of workers who were employed by the independent oil companies, which further complicated the process. Also new blood was constantly added in an environment where educational attainment was not particularly necessary to move from the working-hand level to the lower-management level. When it was all said and done the workers in the oil patch lived in a very egalitarian environment overwhelmingly dominated by wage earners.

Then there was the criminal element, which always had a high visibility in the press. They tended to stay through the height of any particular boom when chaotic conditions allowed them to flourish. The booms only lasted from one to three years and as soon as things began to settle down the criminals usually moved on. Journalists tended to overemphasize the spectacular criminal activity because that is the nature of their business. Consequently the general public tended to group all the boomers with that sort of activity that gave them the reputation of being what eventually came to be called *oilfield trash.* Almost without exception boom participants agreed that gambling, bootlegging, prostitution, robbery, and other assorted criminal behavior was largely confined to specific areas within boomtowns and seldom spilled over into other spaces. The bulk of the hands worked hard, kept to themselves, and seldom had any problem with the bad element or with the law.

CHAPTER 5

# The Panhandle: Populating Cow Country, 1919–1930

When World War I ended in 1919 the Texas Panhandle was much the same as it had been since the turn of the century. Although the land boom that lasted four or five years beginning in 1905 had greatly increased area population, the region still remained sparsely settled. At the end of the war there was not one mile of paved road in any of the Panhandle's twenty-six counties, and cattle ranching remained the mainstay of the economy. However, the oil fever that was inundating the state was gradually being felt in the Panhandle.

The downstate rumblings of big money to be made in "black gold" infected the imagination of M. C. Nobles, an Amarillo grocery man. He caught the fever in 1916 when an employee persuaded him to investigate some Oklahoma oil properties. Ever the cautious businessman, however, Nobles hired pioneer petroleum geologist Charles N. Gould to survey the prospective leases. Gould advised against the deal and, though deeply disappointed, the Amarilloan heeded the warning.

As Nobles was leaving the meeting with Gould he suddenly turned and asked, "Have you ever heard of there being any oil or gas in the Texas Panhandle?" That fateful question was destined to create tremendous changes in the economy and lifestyle of the region. It seems that Gould, recalling a surface survey he had done ten years earlier, did remember a likely oil-bearing geologic structure in the area. Nobles hired the geologist to map the dome, later called the John Ray Dome, which was located about twenty-five or thirty miles north of Amarillo in the breaks north of the Canadian River.[1]

After completing the survey Gould conducted Nobles and a group of local investors to the site and proceeded to give them a remarkable speech that sums up the nature of the oil business of the day. He said:

> Nobody knows whether or not the Lord has put any oil or gas in the Panhandle of Texas. The only way to find out is to drill a hole in the ground. This will cost money. The most likely place I know of to drill this well is right here on this spot. The reason is that this appears to be the highest point of an immense dome. Oil and gas usually occur under similar structures. Because of the size of the structure and because we are to drill near the top, I believe that we are more likely to find gas than oil in the first well.[2]

Nobles immediately formed the Amarillo Oil Company by drawing together a group of local businessmen who invested a small amount of cash. They contracted with an Oklahoma City drilling contractor named C. M. Hapgood to drill the well in return for a percentage of the lease. That initial deal, which involved local people with little or no experience in the oil business using experienced Oklahoma personnel for professional expertise, became a persistent pattern of the Panhandle oil development. The Amarillo Oil Company's Masterson #1 "spudded in" late in 1917 and, after experiencing tremendous difficulties involving all the problems of a "poor boy" operation, the well came in during December of 1918.

The Panhandle discovery was a gasser that produced 15,000,000 cubic feet per day, but there was no market for gas. Nevertheless, the discovery created considerable excitement among experts who were certain that the more lucrative oil would soon be found in association with the abundant gas. The fact that a well drilled in the wilds of northwestern Texas more than two hundred miles from the nearest known production by a group of people who had no idea what they were doing would usually be dubbed rank foolishness. That they drilled a producer was unquestionably seen as some sort of minor miracle in the eyes of the oil fraternity.[3]

Oil hysteria immediately gripped the Panhandle. The February 6, 1919, issue of the *Daily Panhandle* stated that "women and laborers are to be seen on the streets every day with their pockets full of leases and stock forms and a blue print of the lands north of the city under their arms buying and selling stocks and leases." The flurry of excitement caused by the successful completion of the Masterson #1 prompted several major oil companies, notably Gulf, Magnolia, and the Texas Company, to lease a considerable number of large blocks of land. The Amarillo Oil Company continued to drill on its leases and brought in several additional gas wells. The same group of local businessmen who formed the Amarillo Oil Company established the Panhandle Pipeline Company, which built a gathering system to deliver gas to yet another associated company, the Amarillo Gas Company. By the fall of 1920 this associated group of businesses brought natural gas service to Amarillo and soon was able to provide service to a host of other Panhandle towns. The sudden availability of an abundant, inexpensive fuel source to a population that had heretofore imported coal for energy created a small revolution in the area's domestic economy.[4]

One of the unusual things about the Panhandle field's commencement was that wildcat drilling activity was spread out over a large geographic area. Sparsity of population meant that labor was not readily available. The rig owners were therefore forced to set up more or less self-contained operations in the remote areas where they operated. Luckily, the owners were exclusively using cable tool rigs so the crews were small. It took only a driller and a tool dresser to operate a rig. Working twelve-hour tours, four men could keep the rig running twenty-four hours a day. The only additional helpers needed on such an operation were a couple of roustabouts to do general labor such as ditch digging, waterline laying, and a variety of other necessary chores. The last but certainly not the least important member of those operations was a cook. With a seven man crew of that nature housed in two frame shacks, one for sleeping and one for cooking and eating, an isolated rig could be maintained in a reasonably efficient manner.

That was the way oilfield work proceeded immediately following the initial discovery in the Panhandle. Rigs quickly became scattered from as far west as the New Mexico state line to as far east as Wheeler in Wheeler County, just a stone's throw from Oklahoma. Interestingly, the owners had considerable success, primarily in finding gas. Oil was more elusive and no boom developed for a while. Then in the spring of 1921 the Gulf-Burnett #2 well, located a few miles north of the town of Panhandle on the 6666 Ranch in northern Carson County, was brought in. Although the well produced only 175 barrels of oil per day, its existence confirmed the general feeling that oil existed in the Panhandle.

The story of that first oil discovery is also the story of local businessmen getting involved in the oil business. The lease on the Gulf-Burnett #2, held by a group of local investors, was about to lapse. Eugene E. Blasdel, an Amarillo financier, and several of his partners quickly assumed control of the lease, and Mr. Blasdel persuaded the Gulf Oil Corporation to drill on the property in return for a percentage. The Gulf-Burnett #1 was a gas well and somewhat of a disappointment, but the #2 well became the first oil producer in the region. Blasdel was so elated with his success that he named his newborn son James Gulf Blasdel.[5]

Following the discovery of oil in 1921 the pattern of drilling isolated wildcats continued, but the work began to be concentrated in northern Carson and southern Hutchinson counties. There, several more oil producers were soon found. For the next four years that activity increased in tempo although no big strikes were made. By 1925 the outline of the field was beginning to take shape. The Panhandle oil and gas field stretched for more than 120 miles from near Dumas in the northwest to Wheeler County in the southeast. It ranged from five to fifty miles in width with most of the production coming from the 2,500 to 3,200 depth. The field lay along the line of a buried granite ridge named the Amarillo Mountains, a geological feature dividing the field into two distinct sections. The one-third lying north of the Amarillo Mountains was the oil-producing section and the two-thirds that lay to the south produced primarily gas. It was immense, the largest-known gas field in the world at that time, and it drew its match in the men and women of strong character and dogged persistence who eventually developed it.[6]

One of those characters the activity attracted was S. D. "Tex" McIlroy who in many respects represented the stereotype of the adventurous Texas oilman. Born in Hood County, west of Fort Worth, Tex left home at an early age to take part in the Alaskan gold rush of the 1890s. He spent several years in the gold fields where he gained considerable experience in hard rock mining. In the process he acquired a family to support. Although he had adventure enough to satisfy most men bent on experiencing the world, fortune eluded him. Eventually, he packed up his family and traded in the cold north land for a more settled life in Amarillo, Texas, where he went into the grocery business with his brother, White McIlroy.

That is where he was in 1919 when oil fever swept across the barren landscape of the Texas Panhandle. McIlroy could not resist the allure, and once again he was caught up in the excitement and glamour of the search for wealth and excitement. He and his brother formed the Dixon Creek Oil Company, went to Oklahoma where they bought a cable tool rig and hired a drilling crew, and began drilling on leases they held at various spots across the Panhandle. By 1925 Tex was operating on the Smith Ranch in southern Hutchinson County. As midyear approached he had brought in the Smith #1 at a production rate of 300 barrels per day. While not rich, he was a happy man. As work was progressing on that well an incident occurred that illustrates the living circumstance on a wildcat location in a remote area of the Panhandle. Tex contacted his daughter, Hazel, who lived in Amarillo and asked her to bring a load of groceries to the rig where they were running low on supplies. When she arrived at the location, the young lady, who was an excellent cook, discovered that the company cook had quit and that the drilling crew was threatening to leave unless they began getting decent meals. Hazel assumed cooking duties and forever after claimed major responsibility for her father's successful completion of the well.

After the Smith #1 came in as a producer Tex found another cook and work began on the Smith #2. Completed late in 1925 for a production rate of slightly more than 1,000 barrels per day, the second well promised great things for the Dixon Creek Oil Company. Encouraged by his good fortune, Tex decided to deepen the first well in an effort to increase production. On January 26, 1926, he reentered the well, drilled two feet, and stepped back out of the way as the Smith #1 blew in at 10,000 barrels per day. Oil business in the Panhandle picked up.[7]

Within days speculators spent over four million dollars on oil leases. Within weeks thousands of men had flooded into the area. By the end of the year over eight hundred wells had been completed in Hutchinson County. Little Panhandle City, located on the Atchison, Topeka, and Santa Fe Railway about twenty-five miles south of the frenzied activity, benefited tremendously from the boom. For a time in 1926 the local depot handled more freight than any other on the Santa Fe system, with the exception of Chicago. Special sidings were jammed with freight cars loaded with pipe and drilling equipment. For more than two miles along the railroad right-of-way work continued around the clock as teamsters loaded their big eight-wheeled freight wagons and the newfangled motor trucks for the trip to the oilfields. Residents claim that in the first months of 1926 Panhandle

City's one-thousand person population exploded to 10,000. The excited oilmen lived in tents and shacks of every description while waiting to go north to the scene of the action.[8]

Meanwhile, enterprising newcomers developed plans to build several towns at or close to the scene of the drilling. While at least twenty such places were planned and built, most of them, including Gee Whiz, Signal Hill, Whittenburg, and Electric City, lasted for only a relatively short period of time. One, Borger, thrived during the boom and has survived to the present. In its heyday Borger exemplified the traditional stereotype of a lawless oil boomtown. It got its start in the fertile imagination of Asa P. "Ace" Borger, a Carthage, Missouri, native. An experienced town builder, Borger had been involved in the founding of both Picher and Drumright, Oklahoma. On March 8, 1926, only two months after the Smith #1 blew in, Ace established his namesake in the midst of the booming Panhandle oilfield.

Borger had heard of the discovery from his sister who was cooking for one of the wildcatters in the area. He immediately recognized the financial possibilities of a town in the field. Ace paid local rancher, John F. Weatherly, $10,000 for 140 acres and laid out the town on the property. To do this he simply plowed furrows marking off Main Street and planted stakes along the furrow lines to denote lot size. On March 8, 1926, the town site went on the auction block. By sundown Ace Borger reportedly cleared $100,000 and by the end of the year he had grossed $1,000,000 on the sale of town lots.[9]

The thousands who flocked to the new town represented a polyglot assortment of human types. Most were oilfield workers of varied expertise, many were a variety of small businessmen, and a significant number were criminals preying on the uncontrolled crush of humanity. So many came from Oklahoma that the town soon became known to most as "Little Oklahoma," and the popular saying was the town was called Borger during the day and "Booger" at night in keeping with the community's reputation for lawlessness.

R. L. Johnson and his father were typical of the small businessmen who made the Borger boom. They were operating a tiny bakery in Geary, Oklahoma, when they heard of the new discovery. In the late spring of 1926 the Johnsons made a trip to Panhandle City where they spent the night in a lumberyard, the only lodging they could locate. Next day they drove to the Borger town site and were astounded by the magnitude of the activity. After standing in a long line to get a one-dollar plate of pork and beans, they went in search of the real estate office. The only lot left for sale in the new town was located on the extreme south end of Main Street, for which the promoters demanded $750. The elder Johnson refused to pay what he considered such an exorbitant price and the pair started for home. Due to the late hour the travelers were forced to camp beside the Borger-to-Panhandle road for the night. Neither man was able to get any rest

*Newly born Borger in the late spring of 1926, just five or six years following the North Texas booms; automobiles are already dominating the scene and changing the face of transportation in the oil patch. (Photo courtesy Southwest Collection/Special Collections Library/Texas Tech University, Lindsey Nunn Collection)*

due to the noise of the steady stream of cars and trucks that went by all night. By daylight the elder Johnson reconsidered his original decision and decided that perhaps $750 was not too great a business investment to make in a town that had such a large and growing population. Thus, father and son returned to Borger and purchased their lot.

The Johnsons dismantled their Oklahoma operation and shipped it by train to the railhead at Panhandle City. The younger Johnson busied himself constructing the new bakery while his father hauled materials to the site from the railroad. One day while Mr. Johnson was gone, four men wearing sidearms drove up and informed R. L. that they were local lawmen and that they were commandeering his truck to transport confiscated beer into town. Young Johnson went along with them to a remote spot where they raided a sizeable camp that included two large tents. The tents were stacked high with cases of beer, which they quickly loaded on the commandeered truck. Then the lawmen ordered young Johnson to deliver the liquor to a local drug store. He refused that chore and demanded that one of the officers accompany him back to town. They finally agreed and the load of illegal beer was taken to the jail where they unlocked a cell and ordered four or five prisoners to unload the cargo. When R. L. insisted on payment for the use of his truck he was told to shut his mouth or he would wind up in jail himself and probably lose the truck to boot.

Within a week three of the men who had commandeered the truck returned to the bakery. This time they visited with the elder Johnson. The lawmen strongly suggested that the newly opened bakery could easily be robbed without their protection. Mr. Johnson retorted that he would be glad to cooperate with such an obviously needed source of protection, but he felt that he had already made a substantial deposit on protection with the donated truck rental fee that he was yet to receive. Johnson never paid the men for protection and he was never robbed; however, most Borger businesses regularly paid the criminal element for protection during the first year or two of the town's existence.[10]

As the Johnson family's experience indicates, Borger's law enforcement capabilities were less than adequate. Although the town boomed during the midst of national prohibition, there was never any problem purchasing alcoholic beverages within the city limits. Drug stores sold whiskey, home brew could be bought almost anywhere in town, and the rough country of the Canadian River breaks concealed numerous stills and beer fermenting operations of all types and sizes. The criminal element operated the illegal sale of alcohol as a well-organized and efficient enterprise. A group of individuals known as "the line" controlled all liquor sales, and anyone selling such potables in town had to have their approval and pay a percentage to their organization in order to stay in business. All wholesale alcohol was bought from "the line," and they strictly enforced their rules through control of the local law enforcement community.

The north end of Borger was home to the dance halls and houses of prostitution. Perhaps the best known of the dime-a-dance, or taxi joints, was Mattie's, a large barnlike structure with a band at one side and a dance sales ticket booth just inside the entrance. Presiding over the entire enterprise was Mattie Castleberry, who kept order to the proceedings from her throne on a raised platform in the center of the room. Men entering Mattie's bought twenty-five-cent tickets entitling them to dance with any of the scores of girls waiting to entertain. A man would pick out a girl, give her a ticket, and dance a minute or two to a fast-paced tune played by the band. It was not unusual for an inebriated oilfield hand to spend twenty or thirty dollars in one of those establishments in the course of a single night.

Although the dance hall girls did engage in a little freelance prostitution, they were not considered prostitutes by the local residents. The "houses of ill repute" were mostly located on Dixon Street in the same general area as the dance halls. The brothels were usually one story hotel-like structures featuring long hallways lined with tiny rooms on either side. They always had big storefront-type frontages with glass showcases where the working girls could sit in rocking chairs and wave to passers-by and tap on the windows or otherwise try to attract attention. It is estimated that there were as many as two thousand prostitutes working in Borger at the height of the boom.

The much-publicized lawlessness at Borger was the result of the criminal element gaining control early in the development of the town. The man considered largely responsible for the activity was John R. "Johnny" Miller. He acted as Ace Borger's legal advisor and partner in the development of the town site. He also served as Borger's first mayor. Miller was familiar with the wide-open Oklahoma oil boomtowns and consequently felt that the best way to operate such a place was with very few controls. Another factor was that, almost without exception, those who went to Borger planned to be there only long enough to make a little money and then move on to the next boom. This meant that the business element had no interest in building a strong foundation for steady growth in the community. The town's chief law enforcement officer was "Two Gun" Dick Herwig, a man reportedly under indictment for murder in Cromwell, Oklahoma. His primary responsibility seems to have been collecting fees from bootleggers and weekly fines from the prostitutes.[11]

Within a year the general lawlessness combined with a rash of murders prompted Governor Dan Moody to intervene. On April 4, 1927, he sent a company of Texas Rangers into the town to restore order. The rangers ordered several hundred suspected criminals out of town, confiscated and smashed hundreds of slot machines valued at more than $30,000, closed all the beer joints and dance halls, and removed the entire Borger city administration from office. During one memorable raid the officers destroyed 12,000 pints of beer and over the course of their stay they wrecked scores of stills and smashed hundreds of barrels of fermenting mash. Nevertheless, just days after the rangers left, many of the illegal operations were back doing business as usual. By 1929 local efforts at controlling illegal activity were in such turmoil that a crusading district attorney named Johnny Holmes was assassinated by parties unknown. Once again Governor Dan Moody was forced to take a hand, but this time he declared martial law in Hutchinson County, and before the crackdown ended criminal control of Borger was a thing of the past. By 1930 the boom had also run its course and the town settled down to a relatively peaceful and law-abiding existence.

Borger's burgeoning population was the key to attracting small legitimate businessmen as well as the better-known illegal enterprises. Because almost all the businesses stayed open twenty-four hours a day there were hundreds of people bustling along the main thoroughfare at any hour of the day or night. No one knows how many people lived in Borger during 1926 and 1927, though estimates have ranged as high as 25,000 to 35,000 people. So many of the workers lived in camps or isolated work areas and only went into town to spend their money that there was no way to estimate who was actually domiciled within the city limits. The manner in which all those people spent their money is indicative of how they lived, worked, and played in such an isolated part of the state.[12]

Finding a place to live presented a major problem for early Borger residents. The rapid influx of people made it impossible to have adequate housing, just as it had in previous booms, but this time the workers were dealing with a town only months old in a very isolated region. Accordingly, most professional men such as geologists, engineers, and land men sent to the area by the major oil companies lived sixty miles away in Amarillo. They drove to Borger as the job demanded and slept in company-provided facilities while doing their fieldwork. This presented a considerable hardship due to the nature of the drive over a rough, unpaved terrain, but it was a better alternative than living in a shack in Borger.

Lawrence Hagy, who was a land man and an independent oil producer during the Borger boom, was typical of those men. As he remembered it, the road from Amarillo was a dirt track that split occasionally as a side road went to a new drilling location or some lonely ranch house. Because there were no signs marking the way, travelers constantly took wrong turns and got lost. The road was not maintained on any regular basis and during rainy weather, or just because of the dust raised by cars during the long hot summer, drivers moved over and drove parallel to the main traveled road. This practice created a situation where it was not unusual for the roads going into Borger to be a hundred yards or more wide in spots. Hagy said that local ranchers objected that much of their best grazing land was being destroyed because of the uncontrolled driving habits of the oilmen. Depending upon particular circumstance, Hagy usually stayed overnight; but on occasion his trips lasted as long as a week or ten days due to unexpected problems with the fieldwork. While away from Amarillo he slept in one of the shotgun houses he provided for his lease men, but he usually took his meals in town. Larger companies maintaining widely scattered operations usually provided camps complete with bunkhouses and mess halls where their workers lived on a permanent basis. Those facilities also provided temporary accommodations for company men while they were away from their homes in Amarillo.[13]

Business people who flocked to Borger to make some quick money and then move on experienced a markedly different situation. R. L. Johnson, who established his bakery during the first weeks of Borger's existence, said that there was so much work for his family to do that they barely had time to rest. On their first morning of operation, about an hour after the Johnsons put their sign out, a truck stopped and the driver ordered one hundred loaves of bread for a pipeline crew. That was an entire day's production for the Johnson's little establishment. In order to meet the continuing demand they worked in shifts and slept in a lean-to built on the back of the building. Much the same situation held true for Mrs. Claude Ruby and her husband who operated the Nile Drug Store. They remained open twenty-four hours a day and were crowded as late as two or three o'clock in the morning. The Rubys worked ten- or twelve-hour shifts each, slept in the back of the store, and ate at a local café.[14]

*Interior of a typical oil company mess hall where evening chow is being prepared, ca. 1918. (Photo courtesy Tulsa City-County Library, A. I. Leversen Collection)*

Nearly every merchant in Borger responded similarly to the volume of business. Due to the lack of housing and the frantic pace of trade, most of the shopkeepers and their employees slept on the premises. This served the twofold purpose of providing much-needed living quarters and it discouraged burglaries, which were common in the raw oil town.

Not even the postal service was exempt from the overcrowding and thievery. Young Gus Keith went to Borger as a temporary postal clerk in the summer of 1926. He was there for a little over a year and vividly recalled the experience. The mail arrived stacked twelve or fifteen feet high on flatbed trucks. Letters and packages were nearly always addressed to general delivery. The mail clerks would sort as much of the mail as possible and then open the general delivery window to dispense mail to customers who were waiting in lines as much as a block long. Most of the men in the line were picking up mail for as many as six or seven people. For the first few weeks the work kept the clerks so busy that they were always late getting to their rooming house at night. This soon changed when the young postal employees convinced the postmaster that they could prevent mail robberies if they slept in the mail sorting room at the post office. There they put mattresses and springs on the floor for sleeping and rigged up a rope-and-pulley apparatus to hoist the beds to the ceiling during the day. At one stroke the new arrangement foiled constant postal burglary attempts and saved the postal workers the expense of room rent.[15]

The vast majority of the people who went to Borger in 1926 were oilfield workers. These people were roughly divided into two groups: those who worked for the oil companies, both majors and independents, and those who worked for the oilfield contractors. Most of those who worked for the majors lived in company-provided camps. The camps consisted of rows of identical one- and two-bedroom houses for families and one or more large bunkhouses for single men. A camp might be as small as four or five houses or it might include dozens of structures. Some were gigantic, like the Gulf Dial Camp constructed by the Gulf Oil Corporation and the Marland Camp built by the Marland Oil Company, which consisted of almost a hundred houses each. The largest of them all, built by the Phillips Petroleum Company to house the workers at their huge gasoline plant near Borger, contained several hundred buildings. This community assumed all the amenities of a formal town and eventually became Phillips, Texas.

Competition to work for the major companies such as Phillips was fierce, and one of the reasons was the likelihood of getting company housing. Company housing cost the employees a nominal amount of rent, usually in the range of ten to fifteen dollars per month, which included electricity, water, and gas. Houses, streets, and yards were maintained by the company. Naturally, there were some restrictions that tended to rankle tenants. A good example is the way the houses were decorated. They all looked exactly alike on the exterior and the interiors were usually painted a uniform white or beige. One day a lady at a Gulf camp decided that she wanted a nice blue kitchen and proceeded to paint hers that color. When word of the redecorating scheme reached the camp superintendent he sent a crew to the house to change it back to the prescribed color and let the housewife know in no uncertain terms that there would be no deviation from company policy on house colors.

A somewhat similar housing situation developed on a smaller scale among the independent oil operators such as Lawrence Hagy and Tex McIlroy. Their companies built individual houses near their wells. There they could house the lease men they hired to look after one or more wells on individual leases. In order to get dependable family men to work for them the independents offered free living quarters as an inducement. Next to working for a major oil company this was the best situation for an oilfield worker with a family to support.[16]

Those oilfield workers who did not work for oil companies comprised the largest group that went to Borger or any of the other oil boomtowns. These were the men who made up the drilling crews—pipeliners, tank builders, rig builders, roustabouts, and the dozens of other occupations necessary to start up an oilfield—as opposed to the oil company workers whose jobs usually involved operating the field on a long-term basis after the initial drilling and infrastructure was installed. Men who worked for the oilfield contractors were doing most of the general oilfield work for the oil companies. They swelled the town's population and lived anywhere they could get a room. The hands often slept in cars or in company warehouses and more than one spent the night curled up on a pool table, a privilege that cost twenty-five cents per night. During the first weeks of the boom men pitched tents, but as cold weather approached the tents received board floors and wooden walls up to three or four feet on the sides. The vast majority of those men were single, although a number of families did follow the oil booms.[17]

Because of the total lack of housing at Borger the lumber business prospered beyond imagination. Jack Knight and his father bought space in Borger on the day the lots went on sale in 1926. They immediately moved their lumberyard from Davenport, Oklahoma, to the new oil town. Knight remembered that the brothels were their best and most reliable customers because they always paid in cash. Oil companies purchased vast quantities of lumber from the Knights to build their camps, which always consisted of bunkhouses and mess halls first and individual houses afterward. The company camp houses were the most soundly constructed buildings in the area during the boom. The Knight lumberyard specialized in building "shotgun" houses that were the primary living quarters in town. They were simple two- and three-room houses with each room located directly behind the other. The name reputedly comes from the fact that a shotgun fired through the open front door would allow the shot to travel the length of the structure and exit the back door without touching anything. The Knight's basic plan called for a standard ten-foot by twenty-foot, two-room, shotgun house they called the boxcar. It sold for eighty dollars and could be erected in one day. This model, as well as a deluxe ten-foot by thirty-foot house, was replicated by the hundreds in 1926 and 1927 when the boom was at its height. Vacancies were so scarce that these small inadequate structures rented for seventy-five dollars per month.[18]

The folk who resided in those hastily thrown up structures enjoyed few luxuries. There was no electricity, gas, or running water in the town until the late summer of 1926. During those first months businesses and homes were lit by kerosene lamps and water was delivered in barrels. Water, which cost one dollar per fifty-five gallon drum, was delivered from questionable sources on a daily basis. Enterprises like the Johnson Bakery paid as much as ten dollars per day to have enough water to conduct normal business. Cafés charged ten cents a glass for the precious liquid, which some said had to sit for as long as ten or fifteen minutes in order for the sediment to settle and allow the color to become clearer than strong coffee.[19]

Although Borger's streets were unpaved, they did boast good solid two-by-twelve wooden sidewalks. This proved to be very fortunate indeed because the

*Oilfield workers and their families stand in front of two of the most ubiquitous examples of oil boomtown dwellings: the boxcar-style shotgun house and the half-walled tent. (Photo courtesy Oklahoma History Society, Devon/Dunning Petroleum Industry Collection)*

spring and summer of 1926 brought unusually heavy rainfall to the area that turned the town's main street into a quagmire. The constant stream of traffic churned up such a mess that it made the thoroughfare all but impassable. "Chug holes" of such size developed that witnesses swear that more than one heavily laden truck actually overturned in the downtown area. The entire episode presented such a spectacle that an outsider might suspect that there was a complete breakdown of city services; but, as one Borger resident of the time attested, "that could not be true because there were not any city services at all."[20]

Amid the background of all that chaotic and often lawless activity there also existed a solid and public-spirited group. Early in the life of the community several churches were founded by family-oriented individuals among the oilfield workers. The Methodists are credited with establishing the first church in town, followed soon thereafter by the Baptists. The building of the original First Baptist Church in Borger presents an interesting insight into the oilfield mentality of the congregation. One of the volunteers who helped construct the building recalled that the congregation of hard-working and hard-living oilfield workers was sorely pressed to raise enough money to complete the structure. As is often the case with such efforts the members of the congregation solicited assistance from the local business community. One of the first individuals asked to help was the superintendent at the giant Phillips Petroleum Company refinery under construction just outside town. The church deacons asked the superintendent to donate enough cement for them to pour the foundation of their little church and the man flatly refused their plea. He told them that the only way they could get that cement from him was to steal it. "So," related one Borger Baptist, "we did."[21]

While this incident may not necessarily reflect the Christian character of the community it does illustrate the egalitarian attitude of the oilfield hands of the day. Nobody in Borger was particularly wealthy. All Borger residents represented a working class whose highest echelons consisted mainly of small business owners and employees of the major oil companies. They viewed illegal activities as a necessary evil associated with all oil booms and the borrowing of a little cement from a huge corporation for a worthy cause was more in the realm of a loan rather than a theft. They considered the mayhem associated with gambling, bootlegging, and prostitution up on the north end of town as the "real" illegal activity. Most of them had little to do with that segment of the population. Of more immediate concern to the general population was the day-to-day struggle in a raw frontierlike environment.

The women who followed their oil worker husbands to the boom had much to contend with. Everything was in short supply. Water was delivered in barrels, those living in tents cooked outdoors, and on occasion no cooking at all was done because of nearby gas wells that were venting volatile vapors into the atmosphere while being completed. The bountiful gas supply of the Panhandle field also created another impediment to comfortable living. A major outlet for gas usage soon became the manufacture of carbon black. Numerous veterans of the Borger boom remember their first view of the town as one of billowing clouds of black smoke completely obscuring the sky. Borger was ringed by thirteen carbon black plants, and each of them spewed black sooty clouds of carbon into the air. A slight change in wind direction was often the culprit in ruining a snow-white wash that had taken some unlucky housewife all day to complete on a rub board. Carbon black definitely did not make that or any other general housekeeping chore any easier.[22]

Borger represents the archetypical wild and wooly oil boomtown where sheer numbers of residents flooding into a small area created unbridled speculation that temporarily resulted in a vacuum without any particular law and order. Although that situation often existed, particularly when an oil town was developed where no municipality existed before, it was largely a matter of degree. Most booms developed around an existing town that was able to control matters much more effectively. While Borger captured lurid headlines emphasizing the more exciting aspects of life in an oil boom, just thirty miles to the southeast a less spectacular but more economically significant oil development took place around the small town of Pampa in Gray County. During the period between 1926 and 1928, when Borger was enjoying its flush production period, discoveries in the Pampa area went virtually unnoticed. The first finds in that field were made just five miles south of town during 1925. Between then and 1928 various discovery wells were drilled in the Red River breaks southeast of town that opened up the South Pampa, Lefors, and Bowers fields. Those activities gradually gained momentum, which, in 1929, mushroomed into a full-fledged boom.[23]

Comparing oil production figures from the Borger area in Hutchinson County and the Pampa area in Gray County helps to establish the development of the closely associated communities. Hutchinson County boomed in 1926 with production of 24 million barrels of oil. That activity peaked at 33.3 million barrels in 1927, rapidly decreased to 13 million in 1928, 9.4 million in 1929, and 7.4 million in 1930, until it finally settled down in the 3- to 4-million-barrel range during the rest of the 1930s. Gray County by contrast produced almost no oil in 1926 and only 3.7 million barrels in 1927. However, by 1929 the Gray County production exceeded Hutchinson County's by a two-to-one margin and during the 1930s ranged around 10 million barrels per year as compared to Hutchinson's 3 to 4 million. The exact change in the relative position of the two geographic areas can be pinpointed to March of 1929 when Hutchinson County produced 25,200 barrels and Gray County produced 26,400 barrels. From then to the present Pampa and Gray County have consistently out-produced Borger and Hutchinson County.

If all the production numbers illustrate the changing economic fortunes of Borger and Pampa, population figures perhaps help explain the variant social development in the two communities. According to the 1920 U.S. census there were only 721 Hutchinson County residents, and Borger did not even exist. At the same time Gray County had 4,663 citizens and Pampa claimed 987 residents. During the decade between 1920 and 1930, when the oil boom in the region developed, both areas experienced significant growth. Despite the estimates of tens of thousands crowding into Borger, the 1930 census lists only 6,532 residents, while the whole of Hutchinson County could boast only 14,848. Meanwhile, Pampa enjoyed a population advance to 10,470 souls and Gray County contained a population of 22,090. Thus, by 1930 Pampa had surpassed Borger in both oil production and population. It had become the preeminent oil town in the Panhandle of Texas and nobody seemed to have noticed.[24]

When the oil boom struck Pampa in 1929 Gray County possessed the necessary preboom population and institutions to withstand the kinds of social turmoil that had occurred at Borger. Pampa, the Gray County seat, was already a settled agrarian community when the boom hit. The fairly numerous preboom population maintained political control of both the county and the city governments, which in turn kept the situation from getting out of control.

The events that occurred in Pampa during the boom, although similar in outward appearances, bore little resemblance to the basic problems that plagued Borger. Whereas Borger quickly rose to notoriety and developed its economic

*In 1927 the last gasp of oilfield teaming is illustrated by this sixteen-horse team pulling an eight-wheeled, wide-tired, heavy-duty freight wagon loaded with a massive piece of machinery destined for the Phillips gasoline plant at Borger. (Photo courtesy Panhandle-Plains Historical Museum, Canyon, Texas)*

*In 1927, while teams were moving some heavy machinery, the newfangled trucks, hauling pipe and a variety of other materials needed by the oil patch, were already the dominant freighting vehicles of the industry. (Photo courtesy Panhandle-Plains Historical Museum, Canyon, Texas)*

and social base on a wave of social disorder and economic waste, Pampa experienced steady growth prior to the 1929 boom. The more than three years of oil development that occurred prior to the boom gave local developers, oilmen, and law enforcement officers to ample time to prepare for the situation before it occurred. Thus, when oil promoters and workers flooded into the Gray County area, local citizens had the resources in place to control the expected social disorder and economic transformation before the newcomers and their activities got out of hand.

The thousands who "made the Pampa boom" endured many of the same adverse conditions that plagued those who went to Borger. Lack of housing, inadequate municipal services, and a tremendous growth in business activity made life difficult. Fortunately, local law enforcement kept the would-be lawless element under strict control. Naturally there was a certain amount of bootlegging and prostitution, but law enforcement officials never allowed those popular diversions to become an organized criminal conspiracy and create a public spectacle.[25]

As with Borger, Pampa had its satellite towns. Lefors, the best known of these, existed before oil was discovered and became the focus of activity in the Lefors field. Another community, Bowers City, was spawned by the boom when it developed in the middle of the Bowers field. Although Lefors survived the boom period and has since enjoyed a certain amount of prosperity, Bowers City faded from existence within a few years of the end of the boom.[26]

Just as they learned to deal with the social stresses of an oil boom by observing the Borger situation, Gray County residents were also prepared to cope with the economic realities of an oil boom. The Pampa boom began when daily production and drilling sharply accelerated in mid-1929. From January to June of that year daily production rose from 20,000 barrels to 50,000 barrels. Then between June and September production doubled to 100,000 barrels per day. Realizing that a gigantic boom was in the making and wanting to avoid the resulting chaos that enveloped Borger, in August local oilmen established a countywide, voluntary, prorationing or curtailment of production. That system rapidly altered the boom atmosphere and allowed the producers to expand their operations in a more orderly and planned manner over a longer period of time. As a direct result of curtailment Gray County's average daily production dropped about 30 percent in just one month to maintain itself in the 65,000 to 75,000 barrel per day range.[27]

In Gray County a potentially short-term oil boom eased into place as a major element in the larger-scale diversified ranching-farming-oil-based economy. Conversely, in Borger the entire economy revolved around oil production and the Phillips refinery just northwest of town. Thus, in the decades immediately following the discovery of oil Pampa and Gray County's economy, based on the triad of ranching, farming, and oil production, weathered periods of economic upheaval in relatively good condition while Borger and Hutchinson County remained tied to the petroleum industry with its slowly diminishing resources and often wildly fluctuating fortunes.

During the decade of the 1920s the results of oil and gas production became a major factor influencing socioeconomic change in the Texas Panhandle. Between 1920 and 1930 the entire twenty-six-county area increased in population from 114,557 in 1920 to 238,618 in 1930. Leading the way was Hutchinson County with a 1,958 percent increase, closely followed by Gray County with a 374 percent increase. Although it was not located in the oilfields, Amarillo, the largest town in the region, grew from 15,494 in 1920 to 43,107 in 1930. Coupled with that fantastic population explosion was the tremendous impact of petroleum production. By 1930 Panhandle oil production had grown to 30.7 million barrels annually, which amounted to 10 percent of the state's total production at that time.[28]

As the decade of the 1920s ended the Texas Panhandle had been transformed from a sleepy, agrarian region into a more dynamic industry-conscious entity. By the end of the decade most of the hardships created by the rapid and unexpected influx of people into that lightly populated area had been alleviated. The newcomers were absorbed into an expanded economy and stability on the Panhandle. Borger, one of the most infamous of the oil boomtowns, had run its course during that period. The contrast between a newly created boomtown like Borger and a more established boomtown like Pampa and the manner in which both handled similar problems had been clearly illustrated during the course of the decade. Further, the region had learned how to deal with overproduction, a problem that was to become an overriding factor in the East Texas boom that followed hard on the heels of the Panhandle experience.[29]

CHAPTER 6

# East Texas: Changes in the Patch, 1930–1935

As the decade of the 1930s dawned East Texas slumbered quietly in a traditional southern agricultural lifestyle. Its gently rolling, well-watered, and heavily timbered landscape was dotted with hundreds of small farms, many of them owned by African Americans. Area communities rose from the lethargy only on weekends when the town squares became crowded with wagons laden with farm families shopping for their store-bought needs. It was a time when the local communities, places like Kilgore, Longview, Henderson, and Gladewater, struggled with the twin specters of a national economic depression and a serious drought. Some residents went so far as to describe the regional situation as poverty stricken.[1]

Many East Texans viewed the possibility of finding oil in their region as a way out of the looming economic nightmare facing them. Consequently, several exploratory attempts developed there in the 1920s. The area caught the attention of oilmen everywhere when on October 14, 1929, the Pure Oil Company completed a modest producer near the tiny one store crossroads village of Van in east central Van Zandt County. Within ten days promoters erected thirty buildings in the town and oil excitement heightened across the region. The following month the major oil companies of Pure, Sun, Shell, the Texas Company, and Humble consolidated their leases in the Van field and began a cooperative production program operated by Pure. This was the first such unitized operation in the state of Texas.[2]

Meanwhile, the mild interest of the majors prompted a variety of independent oil operators to explore the possibilities of East Texas production. One of those efforts originated in the fertile imagination of experienced wildcatter and oil promoter, C. M. "Dad" Joiner, of Ardmore, Oklahoma. He was aware that some leasing had been done by the larger oil companies and decided to use that as a promotional springboard for his benefit. Beginning in 1925 he began leasing small pieces of land from a number of impoverished farmers until he eventually put together a block of leases that approximated 4,000 acres located between the towns of Overton and Henderson in Rusk County. Operating as the typical "poor boy" outfit, Joiner spudded in on the Daisy Bradford #1 in August of 1927. He spent more than three years trying to complete that well. The wildcatter was forced to abandon the original attempt as well as the Daisy Bradford #2 due to various mechanical difficulties associated with his underpowered and poorly equipped drilling rig. Finally, he proved successful on his third try when the

Daisy Bradford #3 blew in on October 3, 1930, at an estimated flow of 5,200 barrels per day. For East Texas residents a miracle had occurred.[3]

The story of drilling the Joiner well is similar to that of numerous others in the oil business of that day when so many tried almost anything to get at the riches they were sure lay beneath their drill bit. Dad Joiner was the epitome of one of those types. "At a time when the so-called poor boy, the underfinanced operator who drilled for shallow oil with the minimum of resources, was common, Joiner ran the poorest of poor boy operations."[4] After he acquired his leases through a variety of less-than-straightforward means that included utilizing the services of one A. D. "Doc" Lloyd who excelled in pseudogeological sales pitches, he persuaded locals to help build a less-than-adequate wooden derrick. Then he rigged it with a ramshackle secondhand set of rotary drilling machinery. Next he hired an inexperienced driller, used local farm workers to work as roughnecks, and even attempted to keep expenses down by feeding the boiler fire with everything from green wood slabs to used tires.[5]

Joiner spudded in on the Daisy Bradford #1 during August of 1927, which he soon had to abandon at the 1,000-foot level due to a variety of technical difficulties. Then he moved over and started the #2 well, which he lost at 1,437 feet when the crew dropped the casing in the hole and was unable to fish it out. At that point he hired an experienced driller named E. C. Laster who explained the circumstances surrounding starting the drilling of the #3 well.

> So I finally persuaded him to skid the derrick and start a new hole. That was started on May 8, 1929. And we skidded the derrick to the southwest until we broke one of the eight by ten sills. Not having another sill or the money to buy it, we leveled the derrick and started and drilled the well at that point.

Sometimes Laster was paid in cash, sometimes he was promised to be paid in oil leases, and sometimes he did not get paid at all. Many times the farmer/roughnecks walked off the job and refused to return until Joiner paid them their back wages. But on October 3, 1930, they managed to complete the well in the neighborhood of 3,500 feet on a firm basis of false promises, near lies, and downright fraud.[6]

The Joiner well completion quickly escalated into a regional sensation. An excited crowd, estimated at eight thousand, gathered at the well site on Thursday, October 2, to watch the drilling in process. Vendors hawked peanuts and soft drinks to the milling multitude. When nothing happened that day, most returned on Friday to continue the vigil. Late that evening she blew in and a great cheer went up. That weekend oilmen, tourists, and locals rushed to view

*At the beginning of the East Texas boom in 1930 these one-hundred-foot wooden derricks were being utilized, although they were making their last stand. Within a couple of years they would be totally replaced by steel. (Photo courtesy Southwest Collection/Special Collections Library/Texas Tech University, C. C. Rister Collection)*

the show. The seven-mile gravel road from Henderson to the well site was filled with cars creeping along bumper to bumper at five miles per hour. The scene grew so chaotic that every available county law enforcement officer was pressed into service to contain the growing crowd of spectators.[7]

Massive drilling followed immediately and the known size of the field expanded dramatically. By early 1931, as numerous additional wells began to be completed, the Joiner discovery proved to be located on the extreme eastern edge of the field. Major drilling activity tended to concentrate at and around the discovery site as well as at the #1 Lou Della Crim, twelve miles north of the Daisy Bradford, and at the #1 Lathrop, a few miles west of Longview. By mid-1931 all the experts were in agreement that a major new oilfield had been found. The massive discovery extended forty-two miles from north to south and varied from five to twelve miles wide across the counties of Rusk, Cherokee, Upshur, Gregg, and Smith.[8]

Thousands of workers flooded into the area in what seemed like overnight. They came from everywhere looking for work whether they were experienced oilfield workers or not. In those days, at the beginning of the Great Depression, a bright spot of employment shone like a beacon of bright hope to desperate men needing to feed their families. As one preboom resident phrased it, "The whole world was broke and everybody came to Longview looking for work."[9] Once again an oil boom attracted a great population to a relatively lightly populated area of Texas. But this time the mix of workers was even more complex due to the economic pressures of the time, which drew workers from greater distances and more diverse backgrounds.

The surge of workers into the area actually created a surplus of labor. Experienced oilman, L. D. Winfrey, stated that "everybody from outside tried to get in there and get one of those little jobs."[10] That situation made it difficult for even trained oilfield hands to get work. A. C. Hopper is a good example. He was an experienced roughneck from Oklahoma, but he had to make three trips to East Texas before finding a job on a rig. Hopper stated that due to the large number of hands clamoring for work, "You had to find someone you had worked with or knew to get a job." The drilling contractors who had lots of jobs to offer could afford to be choosy about hiring help. During the East Texas boom they even lowered wages to six dollars for a twelve-hour tour.[11]

Among the less-skilled workers the competition was even greater. Each time a job opportunity appeared long lines developed to fill the vacancy. L. D. Winfrey witnessed the hiring of a hundred or fewer men from a crowd of a thousand or more during the building of a small refinery east of Longview. For weeks afterward a crowd of a hundred or so gathered at the site each day in the hope that a few more workers would be hired. In those situations the men who did not get hired simply trudged over to town to stand in the soup line where they could get a free meal. Living like that was discouraging to men desperately needing work. For most of those individuals success meant working sporadically at temporary jobs.[12] One of those faceless thousands looking for work in the East Texas boom was N. L. Field. Field moved his family from a failing West Texas farm in 1932 to look for work in the Overton area. He went for weeks working one or two days a week for thirty-five cents per hour trying to feed his family. After a year of living in that precarious situation he finally landed a steady job. His successful experience was typical of many who made the East Texas boom with no previous experience in the oilfield although many others failed to find permanent employment and left the area in disgust.[13]

The goal of all those men was to find steady employment. Some, through a combination of perseverance and good luck, managed to achieve that goal. And the jobs they most coveted were those with the major oil companies. The most prominent majors in the East Texas boom were Humble, Gulf, Magnolia, the Texas Company, and Sun. The competition to work for them was keen. Harold R. Johnson, like so many others, started a thirty-seven-year career with the Humble Oil and Refining Company during the boom. His story, with variations, is characteristic of that of most of the family men who entered the business during the 1930s in East Texas. Johnson arrived in Kilgore on October 3, 1933, looking for a job with which to support his wife and infant child. A month later he found a job in a clothing store that paid fourteen dollars per week, a garage apartment furnished as living quarters for his family was included in the arrangement. That job ended in January of 1934 and Johnson managed to hire out on a refinery refurbishing job that provided work for two or three days per week at a wage of thirty-five cents per hour. The temporary refinery job lasted only six weeks before construction at the site ended. At that point Johnson began to spend his time standing in long employment lines at the various major oil company offices with no success.

Good fortune finally arrived in the guise of a fellow church member who happened to be the district superintendent for Humble. He told Johnson to come to the Humble hiring office as he had many times in the past, but instead of sitting on the rail out front with scores of other men he was instructed to be near the back door. In that manner he and another fellow were hired on the same day, leaving those out front unaware that two vacancies had been filled. Johnson began by doing general repair work for Humble at a wage of fifty-five cents per hour based on a thirty-six-hour average workweek. After about a year he was laid off in what was termed the "suitcase parade." This was a periodic general layoff designed to get undesirables off the payroll. Those the company wanted back, but were low on the company seniority list, were told to hang around town and they would be put back to work. Sure enough, after waiting for ninety days, Johnson went back on Humble's payroll and spent the rest of his working life with the organization.[14]

Much as the East Texas boom served as a bright national economic spot during the Great Depression years, working for the majors served in the same capacity within the context of the boom. Not only did working for the major companies furnish reliable employment, it also provided better housing opportunities. As in all previous oil booms, hordes of workers descending upon a lightly populated region created severe housing shortages. In order to assure adequate living quarters for their supervisory personnel, the majors, and particularly Humble, secured land upon which they built four- and five-room houses, installed water, electric, gas, and sewer facilities, and put their salaried employees in those company "camps." The official company policy was described as follows:

> After the middle 1930s, the company increased its building of attractive and comfortable camps. It not only supplied more houses for field supervisory personnel, at a monthly rate of $3.00 per room, but it also assumed a measure of responsibility for housing other employees and for providing more community facilities. In or near established communities, non-supervisory employees usually found houses for themselves, but in the company owned camps Humble assigned lots to employees who

wished to build there and gave assistance in planning houses, obtaining materials, and so on.[15]

The hourly employees had a slightly different take on the situation although they supported the company. While salaried personnel enjoyed free housing, those hourly people had to find their own living quarters. Most of them elected to live near the company offices and close to the salaried camp in order to be able to walk to work. Many of them simply went out into the woods and built shotgun houses, one-room shacks, or modified tents by putting in floors and half walls. Even if they went to the trouble to get permission from the landowners to use the land, in most cases it resulted in a rental of only ten to twelve dollars per year. All those structures existed without benefit of running water, gas, electricity, or indoor plumbing. This was not an enviable lifestyle.

By around 1937 most major companies were in the midst of trying to alleviate this situation among their lower-level employees. Humble, at their large facility near Kilgore, is a good example. The firm bought land a short distance from their already existing company camp, divided the land into lots, and proceeded to install gas, water, electricity, and sanitary facilities. Then they invited all their hourly employees to move into the new camp. All the hands had to do was either build a house on the property or move an already existing structure onto one of the lots. Their only expense, like that of the already existing camp, was to pay a small monthly utility fee. The new camp quickly filled up although some employees remained outside the company domain just as they were accustomed to doing. Those who moved to this new location soon dubbed it the "poor boy camp" and referred to the location where the salaried personnel still resided as the "company camp."[16]

Those hands not so fortunate as to be employed by majors had to continue as best they could under poor circumstance. At first many of them slept in fields, using old newspapers for cover during good weather, and when it rained or turned cold they sought shelter in churches, abandoned barns, or whatever was available. Those people, like so many working for the majors, also tended to become squatters in the East Texas woods. Their little shotgun houses, tents, and shacks were scattered everywhere throughout the countryside. One place in particular became known as living quarters for destitute people. Known as "Happy Hollow," it was a low-lying area on the edge of Kilgore. Happy Hollow consisted of a motley collection of cardboard houses, scrap lumber shacks, and every other conceivable type of living quarters imaginable. At any given time between one hundred and three hundred people called the place home. It was raided on a regular basis by local law enforcement officers who always managed to arrest a hundred or more citizens on each occasion. Nevertheless, by the next

*During the East Texas boom of the 1930s Humble and other majors began providing camp housing to their employees, like those shown here, in an effort to get and keep reliable employees. (Photo courtesy Southwest Collection/Special Collections Library/Texas Tech University, Jack Nolan Collection)*

day or so it would refill with desperate people in need of work. Long after the boom ended and the memories of Happy Hollow had faded the location became a part of the Kilgore City Park.[17]

Although the housing situation during the East Texas boom had its difficulties, food acquisition proved to be less of a problem. When the boom struck there was only one dining establishment in Kilgore. The City Café, right across the street from the railroad depot, was owned by Crown Dixon; the café and its associated meat market did just enough business to provide Dixon with a modest living. When the oil boom developed all that changed. Within the first four months of 1931 there were 143 eating facilities in the town of Kilgore.

From January 1, 1931, until he sold the business in May of that same year Mr. Dixon did not close the doors the City Café. Men crowded into the place out of the muddy street twenty-four hours a day. The mud they tracked in on their fancy lace-up leather boots and the more mundane rubber work boots accumulated two inches deep on the floor. It created such a mess that the more than thirty newly hired employees only made a half-hearted attempt at scraping it out. As the boom escalated and more and more demand arose numerous hamburger stands and chili parlors cropped up everywhere, but very few full-fledged cafés developed.

The food served at the City Café was typical of other local restaurants. Chili, made up in number three washtubs, was a mainstay. The menu also featured ham and eggs, coffee, and a variety of plate lunches that consisted of roast beef or fried chicken, mashed potatoes, another vegetable, and a dessert. Those plate lunches sold for thirty-five cents.[18]

The cafés of East Texas catered to the oilfield hands in several ways other than providing food at a reasonable cost. One of those activities involved selling meal tickets as a convenience to the workers. When Mary Florey Love worked for the City Café in Kilgore she punched as many as five hundred meal tickets per day for the hands working in the field. The tickets were used in lieu of cash due to the men's unusual work hours, which prevented them from getting to the bank to cash their checks during the day. Additionally, café personnel made regular bank deposits for oilfield workers because of this same inconvenience. Thus, in addition to serving as a place to eat, the cafés became clearinghouses for workers needing to get their financial situations in order.[19]

Another location to secure food and lodging appeared in the large number of boardinghouses and rooming houses that abounded in all the East Texas boomtowns. Those varied somewhat in their character. They appear to have been rather equally divided between the boardinghouses that provided rooms and meals and the rooming houses that provided only rooms. The oil companies, in the first days of the boom, provided bunkhouses along with boardinghouses for their large construction crews. Those facilities, located out in the field and away from town, provided family-style meals for their employees. Similar establishments in town provided family-style meals at a price averaging $1.50 per day. Gerald Lynch stayed at the Como Hotel during 1932 where he had a room and dined in a family-style dining room for the sum of ten dollars per week.[20]

Obviously, the rent and food costs at those East Texas boomtown establishments were modest. By the same token the clientele did not consist of the most fastidious individuals. One of the boardinghouses was located beside a gambling establishment that operated twenty-four hours a day. Many of the gamblers took their meals there alongside hundreds of hungry oilfield workers. On one occasion, after one of the workers had left greasy fingerprints on several slices of bread stacked on a platter, one of the neatly dressed gamblers seated across the table from him eyed the stack and asked for some of the bread with the stipulation that they, "Burn the top two and deal me one."[21]

The many family groups drawn to the area also had to make ends meet under the twin circumstance of oil boom conditions and an ongoing economic depression. They found food costs to be very reasonable. Most working families could be well fed on as little as twenty to thirty dollars per month. This included a good supply of meat and plenty of milk, which sold for five cents per quart. If food was inexpensive and plentiful, potable drinking water was not. Uncontaminated water wells were rare and the rapid influx of such a large population tended to destroy those springs and water wells already in existence. Almost everybody received their water from water wagons that got their supply from any available source. In this area of survival, as in many others, the major oil companies assisted by drilling good water wells and allowing water vendors to fill their wagons at those locations. It was not unusual to see long lines of horse-drawn water wagons standing at those wells from daylight until late at night. Regardless from whence their water came, boom residents all strained and boiled their drinking water before consuming it.[22]

Once food and shelter was taken care of the younger and more transient population demanded entertainment. As in all oil booms this was available in abundant quantities. That ever-present hovering crowd of semilegal characters who always cropped up with the appearance of big oil money left the dying booms of the Texas Panhandle, Oklahoma, and the Permian Basin for more lucrative fields of endeavor in East Texas. Everybody agreed that the new situation was not nearly as rough and rowdy as that of Borger in the Panhandle or Wink in the Permian Basin, but there was a significant amount of bootlegging, gambling, and prostitution. Kilgore stands out as probably the toughest of the East Texas boomtowns, although there was illegal activity enough to go around throughout the field. Probably the existence of well-established communities and a population that did not welcome the rowdy boomers with open arms had much to do with this more subdued atmosphere.[23]

Typical of those early arrivals catering to entertainment needs in East Texas was Mattie Castleberry, who closed her dance hall in Borger and headed for the new promised land. The lowering price of oil and resulting decrease in activity in the Panhandle had decimated her dance hall business there. She set up her new taxi dance hall, which she dubbed Mattie's Ballroom, on the road between Kilgore and Longview to take full advantage of business from both communities. Before long her establishment became the best-known dance emporium in East Texas. Described as "a lady of uncertain years" by Gerald Lynch, Mattie reigned supreme over a domain that provided music and dance partners from around sundown until the wee hours of the morning to a large and enthusiastic audience. Patrons provided their own beverages and paid a modest fee for ice and setups to prepare their drinks. Dances with the girls employed there cost ten cents each for five minutes or less; fast-paced tunes were played by a live band. It didn't take a roughneck long to go through a five dollar roll of dance tickets when he was having such a good time. Mattie's became so well known for its entertainment in a relatively safe environment that many locals frequented the place with their dates or wives to dance to live music and enjoy a liberal dose of the illicit libations proscribed at that time by national prohibition.[24]

As the new residents struggled under difficult conditions simply to survive, the boom continued to escalate. By the end of 1931 East Texas had more than 3,700 operating wells, and during the following year that number soared to 5,652.

When the Joiner well came in oil was selling for $1.10 per barrel, but the massive production caused the price to tumble until by the summer of 1931 it hit a low of ten cents per barrel. On August 15, 1931, the Texas governor, Ross Sterling, declared martial law in the East Texas oilfield and placed troops there to control profligate production. The move was designed to control production and conserve the oil while at the same time raising the price of oil. Although his strategy was succeeding, that December several small recalcitrant producers managed to get an injunction to remove martial law. To circumvent this maneuver the governor simply transformed the troops into law enforcement officers and continued trying to control the wasteful practices prevalent in much of the area's oil-producing community. Although this confrontation created considerable confusion and a certain amount of violence in the East Texas oilfield at the time it marks the beginning of serious oil conservation efforts in the state of Texas.[25]

The turbulent legal situation in East Texas affected workers by making jobs even harder to find. This in turn generated hard feelings toward the concept of oil regulation and considerable resentment against the major oil companies; the small independent oil companies and many workers blamed them for the financial mess. Those problems sparked some violence in the area; several incidents of dynamite explosions on pipelines and other examples of sabotage became frequent. Also state oil production inspectors received almost daily threats and many of them resorted to wearing sidearms. For a time the situation remained extremely volatile, but by 1935 most of the excitement had abated.[26]

The massive overproduction in the East Texas field, due in part to the inexpensive cost of completing wells, attracted hundreds of independent operators who were forced to drill and produce their leases as fast as possible in order to make enough money to stay in business. Utilizing the equipment of the day it took less than a week to drill to the pay zone in the Woodbine sand at a depth of 3,500 to 3,600 feet. The resulting East Texas "sweet" crude oil that was very low in sulfur content found a ready market; combined with the short distance to the refineries on the Gulf Coast this greatly reduced the cost of producing and transporting the oil. All this came together to create a situation where a profit could be made despite the rapidly falling price of oil. This combination of events, which led to the state of Texas intervention, also created a serious conflict between the hundreds of independents and the relatively small number of majors. Many of the independents believed that the intervention was a plot hatched up by the majors in the guise of proration of resources to force them out of business.[27]

Against this larger financial and legal background the process of drilling oil wells proceeded at a fast pace. During the first year of the three-year active duration of the boom, when so much work was done by the hundreds of small operators, it was a common practice to utilize the inexpensive local timber resources to build wooden derricks, which were already well on their way to becoming obsolete in the industry. But that quickly changed as the majors became involved and ultimately the East Texas field was largely drilled with steel derricks. In those days, before weight indicators became common on drilling rigs, many of the old-time drillers flatly refused to work under steel derricks; they believed they were unsafe because under strain they would fail without warning, whereas the driller could tell from the creaks and groans of the wooden derricks just how much strain was being put on the framework and therefore would not overload the capacity of the derrick and "pull the rig in," that is, collapse the derrick. Those traditional drillers were convinced that they could judge how much strain the wooden derricks were under by what they called the "squat," or the distance the derrick compressed when under great strain. They even devised a method of measuring the "squat" by suspending a piece of pipe in the derrick that ended a foot above the drilling floor. By watching that pipe they could judge the amount of "squat" the derrick could stand, which was generally considered to be about eight inches. However, when it was all said and done the East Texas boom marks one of the first times in Texas that a major field was drilled using steel derricks as a primary drilling tool.[28]

The East Texas field became well known as one of the first major discoveries since the Gulf Coast boom of 1900–1905 that was drilled primarily by the rotary method in a hard formation area. There was a general prejudice in the oil business against using the rotary drilling method. Most operators agreed that the method worked fine in the unconsolidated formations along the Gulf Coast, but it did not do well in the harder formations elsewhere in the state. The primary reason for this was the slow drilling of the rotaries in hard formations due to the nature of their bits. Originally they used a drag or "fishtail" bit shaped like a chisel with the face split and half of it facing two ways, somewhat like the tail of a fish. These bits simply scraped their way though the earth and were not the most efficient tool in the world. Drilling efficiency increased somewhat after the introduction of the Hughes roller cone bit in 1909, but its two-cone construction still drilled slowly in the hard formations when compared to the traditional cable tool method. However, over the years the Hughes Tool Company and other manufacturers of similar roller-cone-type bits improved their products until by 1930 they were being generally accepted as comparable to the cable tools. Then in 1933 engineers at the Hughes Tool Company invented the tri-cone bit that so enhanced the drilling speed and accuracy of the rotary method that it soon eclipsed the cable tool method; the use of that style of bit remained the primary drilling bit throughout the remaining twentieth century.[29]

Another argument offered by the traditionalists against rotary drilling involved the mud used in the rotary drilling method. They argued that it did not work well in the harder formations and that it did not give accurate information concerning the formations being drilled. However, their most vehement objection was the claim that the mud had a tendency to seal off the formations being drilled through; thus preventing many of them from producing and even obscuring the smaller pay zones from even being discovered. This was such a hotly debated question that in 1931 the American Petroleum Institute formulated a study that concluded that all the traditional arguments against the use of mud in drilling were unfounded.[30]

In the thirty years between the Spindletop discovery and the development of the East Texas field a considerable cadre of experienced rotary oil well drillers had developed. These younger men who naturally gravitated to the East Texas discovery in a time when work was generally scarce seem to have been more open to many of the technological changes taking place in drilling oil wells. By the same token the older cable tool men were losing their influence on the industry. All this technological change was having a ripple effect on the rest of the industry.

One of the trades definitely affected was rig building. The changeover from wooden to steel derricks created a profound difference in the way derricks were built and in their permanence. Traditionally, some of the wooden drilling derricks had been salvaged and rebuilt for drilling and some of them had been skidded intact to new locations, but the bulk of those wooden structures tended to be left in place to be utilized for pulling rods, tubing, and casing. This practice resulted in the traditional look of the early oilfields: they were dominated by a forest of derricks. Ultimately, this created an unsafe situation when older deteriorated wooden derricks began to collapse on a regular basis. A good example was the 1915 Gulf Coast storm that destroyed all the standing wooden derricks at Spindletop and the May 12, 1919, storm that destroyed 409 derricks at Goose Bay near Houston.

With the general acceptance of the steel derricks the rig builders had to develop a whole new set of skills. These structures were bolted together in a prescribed manner as described in a set of blueprints, and there was little or no room for variation. Further, the steel derricks were much easier to move intact from one location to another, and skidding rigs became a major portion of the work performed by rig builders. Very few of those drilling derricks were left in place. They were dismantled and rebuilt time after time as new drill sites became available. For a time a much lighter pulling derrick replaced the skidded drilling derricks, but over a period of years those also became obsolete as portable pulling units with retractable masts were developed.[31]

The process of skidding a rig in the East Texas field was a colorful sight. The rig builders laid down 3-inch by 12-inch by 30-foot mat boards to create a solid base on which to move the rig. Then they jacked the equipment up and placed it on wooden rollers. The rig was pulled by a truck or caterpillar tractor, or if it was extremely muddy they used teams of draft mules. Once the rig started moving

*Within two years after the beginning of the East Texas boom, wooden derricks disappeared, as illustrated by this 1932 view of Kilgore, forested with steel derricks. (Photo courtesy Southwest Collection/Special Collections Library/Texas Tech University, C. C. Rister Collection)*

the rig builders rushed to remove the rollers from the back of the moving equipment to the front to keep it rolling. Additionally, they removed the thirty-foot matting boards from the back and placed them in front of the slowly moving equipment to assure a solid route for the move. It was hard, backbreaking labor and if for some reason the process ever faltered it was next to impossible to get the rig moving again. Consequently, on rig moving day it appeared to be a situation of mass confusion with everybody involved rushing around like mad, cursing, bellowing orders, and generally doing everything in their power to make it happen. During good weather a move of a quarter of a mile or less took about five or six hours, but in wet weather it would take from daylight to dark to get the job done.[32]

The use of horses to skid derricks during inclement weather in East Texas points out one of the limited uses left to teamsters in the oilfields. By 1930 horses pulling freight wagons were used in a very limited manner, although teams were still very much in evidence in the earthmoving trade, such as the excavation of mud, storage, and disposal pits. Despite the fact that by 1930 there were only 7,300 miles of paved road in the entire state of Texas, the automobile was fast becoming the transportation mode of choice. As early as 1922 there were more than 500,000 cars and trucks registered in the state, and that number was growing at a tremendous rate. This sounded the death knell for the work of the teamsters in the oil patch.

One of the best-remembered aspects of the employment of horses during that period was their use to pull motorized vehicles out of boggy places. The locals made good money by stationing teams of horses or mules at the impassable places on the inadequate roads and charging a hefty fee to extricate the underpowered cars and trucks from the sticky mud. This practice soon led to the development of one of the more common elements in the East Texas oil patch, called "tolls." The tolls dotted the landscape where landowners paved boggy places with logs and charged a fee to use the improved portion of the road. With the passing of the teamsters, one of the more colorful aspects of the early oil industry disappeared as the massive camps of horses and mules and hundreds of teamsters passed from the scene. It seems ironic somehow that the very industry that provided so much employment for that traditional occupation also caused its demise by providing the fuel that allowed the internal combustion engine to become the mainstay of transportation.[33]

In many ways the East Texas boom was different from previous Texas discoveries. It is well documented how profligate overproduction during the boom created serious public and legal consideration toward the idea of conservation of oil and gas production, but it also had several characteristics that directly affected the workers in the field. The boom occurred during a period of massive national economic upheaval and general unemployment. For the first time in an oil boom in Texas there was no labor shortage. Indeed, there was such a surplus of workers that the drilling contractors actually began to cut wages, much to the dismay of the experienced hands. However, that was somewhat compensated for by the fact that the cost of living did not escalate to unreasonable heights. In general housing was just as bad as in preceding booms, but it marked a time when the majors began serious efforts to alleviate those conditions for their employees. East Texas had its share of lawlessness and illegal activity, but it seemed to fall more into the paradigm of the more settled towns keeping that activity under reasonable control. In those respects the workers making the East Texas boom experienced the typical oil boom with modifications.

Of more concern to the "boomer" population who followed the opening of new oilfields across the state, was the realization that it was a time when change was afoot in the oilfield. The conflict between the rotary and the cable tool drillers reached its end for all practical purposes in the 1930s when the East Texas field was drilled by the rotary method. The various technical innovations leading up to that time assured the rotary men ascendancy in the future. It also marked the end of the use of wooden derricks in favor of steel, which in turn completely changed the type of work performed by rig builders and started the beginning of the end of their craft, which would totally disappear over the next couple of decades. The growth of the use of the motor truck, which first saw significant use in the Wichita Falls area some ten years earlier, reached a certain maturity during the East Texas boom, and the teamster trade in the oilfield soon became a thing of the past.

East Texas provides a glimpse of things to come in the oil patch. It represents a time when scientific and technological innovation became much more important in the petroleum business. That in turn put many traditional oilfield occupations on the path to significant change and in some cases total obsolescence. Those technological changes, particularly in the area of transportation, were to have a profound influence upon the lifestyle of the boomers in the future.

CHAPTER 7

# Way Out West, 1923–1940

The Permian Basin is a region that defies the usual practice of being noted for one particular oil discovery or boom. It is the best example in Texas of a single geologic oil-producing feature that includes numerous oilfields discovered over a long period of time in a large geographic area. The Permian Basin is a gigantic area in West Texas that extends some 300 miles in length from near Lubbock on the north to Pecos County on the south and approximately 250 miles wide from near Colorado City on the east to the Pecos River Valley and eastern New Mexico on the west. Geologically it is a deep subsurface basin formed during the Permian period that consists of very thick sedimentary rock formations where petroleum exists at various locations across its vast expanse. Between the 1920s and the 1950s numerous oilfields were discovered in the Permian Basin that provide an excellent illustration of a changing technology and an oilfield lifestyle that was tailored to a specific geographic situation.[1]

The first discoveries of oil in the region were near Colorado City in 1918, but it was not until 1923 that serious activity developed. On May 28 of that year the Santa Rita #1 came in a few miles west of Big Lake and activity began in earnest. The story of the drilling of the Santa Rita well illustrates the pattern that became common for future activities in that arid and sparsely settled ranching region of West Texas where it was necessary to develop self-contained operations far from the nearest supply areas.

It all started when Rupert T. Ricker of Big Lake, Texas, made an application for oil prospecting on University of Texas land in his part of the state. Unable to raise the necessary funds to pay the leasing fee for the project, Ricker sold the rights to Frank Pickrell and Haymon Krupp of El Paso for the sum of $2,500. Then, in April of 1919, the two partners formed the Texon Oil and Land Company as a vehicle for promoting the sale of leases. The scheme took so long to develop due to poor sales that they changed their approach to selling "certificates of interest" in future production. With the looming disaster of the approaching deadline for the lapse of their arrangement with the state of Texas the promoters, who never had the remotest intention of doing any real work on the land, were reduced to actually drilling a well on their optioned property. That was the only way they could renew the arrangement and continue their last-ditch effort to boost interest in the project.

Accordingly, Pickrell hired a geologist to locate a likely spot on which to drill a well. Then he went against that professional advice when he decided that the location chosen by the geologist was so isolated that transportation costs would be prohibitive. He changed the drilling location to a spot 174 feet north of the Kansas City, Mexico, and Orient Railway near Best, Texas, about twelve or fourteen miles west of Big Lake. With time of the essence the promoter managed to transport a portable drilling machine to the location the afternoon of January 8, 1921, and before midnight, when the lease was set to expire, he had spudded in on the water well designed to supply the rig boilers with water. Needing legal proof that he was actually drilling, Pickrell hailed a pair of passing local citizens and persuaded one of them to sign an affidavit that he had witnessed the event. The lease was saved and the partners had a three-year extension on the property.[2]

The high cost and difficulty of transport in the region was the primary reason Pickrell had for ignoring the geologist's report for the site of the well. At that point in time the few roads in the Permian Basin region were little more than dirt tracks, which made travel a slow and laborious process. Unlike the areas to the east and south where previous oil booms occurred, loose sand rather than mud was the primary difficulty. For example, in 1925 it took Harold Morely more than nine hours to travel the 150 miles from Midland to Abilene over what was considered a good graded gravel road. He described another trip from Hobbs, New Mexico, to Big Lake, Texas:

> On the way down there we had one spare tire and we went from Hobbs across the desert to Midland on those sand roads and I think we spent about half the time pulling mesquite thorns out of the tires. It took a long time because we had trouble keeping those five tires full of air. Well, it was a desert. When you got away from a settlement it was sand and mesquite. Of course there were cattle trails, but you could go down a road and you would have a turn—there would be a fork in the road. We finally decided that they always came back together. One road went around the hill this way and the other went that way, but they came back together so it got so it didn't make any difference which fork we took.[3]

Travel by automobile anywhere across that vast ranching country required two people. One did the driving while the other helped fix the numerous flat tires engendered by the ubiquitous thorny vegetation, although that other person mainly served to dig the car out of the seemingly endless series of soft sand beds. Consequently, in those first years of the region's oil development the railroad was the primary mode of transporting freight to the oilfields. From the railroad sidings both trucks and wagons transported the supplies to the drilling locations, but wagons and teams were preferred to haul loads to the actual scene of drilling. Trucks were used more to get supplies from the siding to the supply warehouses because their low horsepower at that time prohibited their efficient use in the sandy country.[4]

Another factor that Pickrell had for ignoring the site choice was that he operated in an era when many oilmen were skeptical of geologists' reliability in predicting favorable well locations. The nature of the Permian Basin petroleum producing area was such that the oil-producing formations were buried so deep that there were usually few surface geologic indicators. Consequently, professional petroleum geologists were destined to wait a few years before they became an important factor in finding oil in the region. As more and more wells were drilled and more subsurface information became available the reliability of finding suitable drilling sites became more precise as geologists interpreted that data. The large oil companies recognized that and during the early to mid-1920s they began to develop significant geology departments. However, for a number of years there remained lingering doubts in the minds of the working hands concerning the reliability of the rock hounds' information. William Allman, who worked as a driller during the 1920s, summed it up for the hands when he stated in 1970, "Well they didn't think too much of them in the early days. It was just if you believed it was there, just drilled a well, if you had the money to do it. I still favor that some." Arthur Stout, another early-day driller, went even further when he commented on the nature of those pioneer drillers: "His attitude was that he didn't want to work with geologists in the first place; he didn't want to have anything to do with them."[5]

In June of 1921 Pickrell hired a rig builder who brought a crew to the remote site where they camped in tents while building the derrick and two boxcar-type shotgun houses designed for the drilling crew's living quarters. Next he hired an experienced oilfield man named Carl Cromwell to act as driller on the job. Cromwell was one of those hard-bitten Pennsylvania-trained cable tool drillers who had started in the business as a teenager and had worked in Oklahoma before making the Burkburnett and Ranger booms. He moved his wife and small daughter to the location in the summer of 1921 and immediately began to install the drilling equipment scattered around the site where the teamsters had left it. Cromwell and his family lived in one of the houses and the tool dresser lived in the other. The isolation and loneliness of the location made it difficult to get and even more difficult to keep good hands. As a result he went through several tool dressers and even used local cowboys from time to time in his ongoing effort to complete the project.[6]

Just prior to spudding in on the well on August 17, 1921, Pickrell climbed the derrick, scattered some dried rose petals in the air, and dedicated the well as the Santa Rita. That unusual event was the result of a promise he made to a group of Catholic women from New York City who had purchased some of the Texon certificates. Concerned that theirs might be a poor investment they consulted their parish priest who advised them to put their trust in Santa Rita, the patron saint of the impossible. Then the priest blessed a red rose in the name of Santa

Rita and the ladies presented the rose to Pickrell with the request that he take it to the location of the well and scatter the petals from atop the derrick as a way of blessing the project. In Pickrell's own words:

> The name Santa Rita originated in New York. Some of the stock salesmen had engaged a group of Catholic women to invest in the Group 1 stock. These women were a little worried about the wisdom of their investment and consulted with their priest. He apparently was also somewhat skeptical and suggested to the women that they invoke the help of Santa Rita who was the Patron Saint of the Impossible! As I was leaving New York on one of my trips to the field two of these women handed me a sealed envelope and told me that the envelope contained a red rose that had been blessed by the priest in the name of the saint. The women asked me to take the rose back to Texon with me and climb to the top of the derrick and to scatter the rose petals, which I did.[7]

A few months later, in January of 1922, an experienced cable tool man named Dee Locklin, who was returning from a drilling job out in Loving County, stopped at the site when he saw the rig running. Cromwell hired him on the spot as tool dresser and Locklin brought his wife out to the site where they occupied the spare shotgun house. Locklin stayed on the job until they finished the well eighteen months later. He remembered waiting for as long as four or five months between paychecks and having a problem securing enough labor to keep the rig operating twenty-four hours a day. From time to time they had to shut the operation down due to a lack of supplies, for the downhole fishing jobs that continually plagued the operation, or for any number of other reasons associated with a "poor boy" operation. According to Locklin's account the rig was idle almost as much as it was operating during the entire well-drilling process.[8]

Finally, on the afternoon of May 27, 1923, at a depth of 3,050 feet, the drilling crew got a showing of gas and oil and shut the drilling operation down. Next morning at approximately 7:00 A.M., while the driller and the "toolie" were having breakfast in their respective houses, they heard a loud roaring noise from the direction of the well. Rushing outside, the men witnessed oil pulsing from the wellhead. It ebbed and flowed several times until it eventually rose high enough to blow over the top of the derrick. The Santa Rita #1 had just blown in for a production of between thirty and one hundred barrels of oil per day. It took a couple of weeks to get the well under control. About every twelve to fourteen hours it would pulse and blow over the top of the derrick then subside to a trickle. The crew filled the two 250-barrel tanks at the location with the production. They even let it run into their slush pit until it overflowed, but the bulk of the oil blew over the surrounding countryside before they finally managed to get enough storage and were able to control the flow.[9]

In the typical entrepreneurial spirit of the day both Cromwell and Locklin managed to profit from the situation. According to Locklin they devised their plan on the afternoon they got the first showing of oil.

> We closed down and nailed the rig up, nailed the door down—the trap door to the cellar—set the bailer in the hole and we were going to not work the next day, we were going to buy some leases around the country, which we did. We did that because we knew we had oil; we didn't know how much, but we actually knew that much before the well began to blow. Then the well blew in. There wasn't anything we could do about it. We just had to go on. We got 20 or 30 sections leased, I can't remember for sure. We got it all in one day. It was quite a bit of country to cover, but we made all the ranches and made the deal right there: then we went over to the courthouse which was at Stiles at that time and had the papers drawn up.

Despite their quick work at cashing in on the situation both the driller and the tool dresser on the Santa Rita #1 never got any production on the leases, although they did manage to dispose of them for a nice profit.[10]

Although he had completed a successful oil well, Pickrell had difficulty getting any of the big oil companies involved in the operation. It was not until October of that year that Pennsylvania wildcatter Mike Benedum decided to take a chance on the isolated discovery. He bought into the operation, formed the Big Lake Oil Company, and began an extensive drilling operation. Within a year the Permian Basin's first big oilfield was a reality, although shipping the oil to market from such a remote area was a major problem. The only means of transport was either by truck or railway tank car and both of those proved totally inadequate. Then in October of 1924 they negotiated a deal with the Marland Oil Company, which brought other major oil companies in on a scheme to build a 400-mile-long pipeline to Texas City on the Gulf Coast. They also established a tank farm in the midst of the field that consisted of seventeen 80,000-barrel tanks. By that time the boom was on and the region was flooded with drilling crews, rig builders, tank builders, pipeliners, and all the others that followed oil booms.[11]

As in all the previous booms across Texas the sudden influx of workers into the area created a major housing crisis. In this case it was particularly acute due to the isolated location of the activity, where absolutely no towns existed. At first most of the company men stayed at Big Lake, which was the only town within a reasonable distance of the field. There was only one small hotel in Big

Lake so H. A. Hedberg and Paul Doran had a hotel in Fort Stockton dismantled, moved it there, and rebuilt the structure. That fifty-room two-storied building filled so quickly that they double-rented rooms and even had to turn customers away. Business was so good that within months they built a second hotel in the tiny town and it also remained filled to capacity. Meanwhile, Best, which was little more than a railroad siding in the heart of the new field, began to develop as the center for the rougher element of laborers who worked for the various contractors. According to H. A. Hedberg, "It was a smaller town and it consisted of mostly oilfield workers and the type of people who follow the boom—boomer types, racketeers, dope peddlers, bootleggers, and what have you." This prompted the two entrepreneurs to build a third hotel at that site. Those three establishments operated at full capacity for about three years before the boom subsided and the original building was once again dismantled and moved to Pyote to be rebuilt where another boom was developing.[12]

With the urgent need for housing their employees and with Best developing as a wild and wooly oilfield boomtown, officers of the Big Lake Oil Company decided to build their own town. Between 1924 and 1926 Levi Smith, president of the company, planned and built Texon just a few miles away as a safe and pleasant place for his employees to live. Texon was widely touted by newspapers across West Texas and beyond as a model town with a hospital, a school, a theater, swimming pool, and several private businesses housed in company buildings. It was so much better than any of the boomtowns that it only seemed a model town. In reality at one end of the town the company built houses complete with numerous amenities that accommodated all the supervisory personnel. At the other end of town the lower-level workers lived in considerably less splendor and with a much lower level of comfort.[13]

While all this was going on a tremendous amount of drilling developed at some very remote locations across the region. Typically, a drilling location would contain a bunkhouse, a cook shack, and possibly an office, although many times that designation consisted of nothing more than a cluster of tents. Water was supplied from a water well, a rancher's windmill, or even freighted in from some other location. The drilling crew was usually hired for the duration of the drilling job in an effort to keep the crew on the job and assure a labor force until the well was completed. They did not receive their pay until the job was finished some three to six months later.[14]

The activity surrounding the development around the Santa Rita discovery soon began to attract the attention of oilmen from across the country. In 1925 George B. McCamey and J. P. Johnson brought in a well about fifty miles to the west of the initial discovery, which precipitated another boom. Town site promoters immediately became very active throughout the region. Within months the town of McCamey developed, first only as a railroad siding where George

*A view of the model oil company town of Texon, which was plunked down in the midst of the first oilfield discovered in the Permian Basin. (Photo courtesy The Petroleum Museum, Midland, Texas, Berte R. Haigh Collection)*

McCamey received his supplies and business was done out of boxcars, then as a full-fledged town. The following year the town of Crane, some twenty miles north of McCamey, was established as a workingman's town amid the Gulf Oil Company's expanding operation, and that same year, 1926, Iraan, twenty miles to the south of McCamey developed in the midst of the newly discovered Yates field.[15]

All of this activity developed in a region so sparsely populated as to be almost deserted. Population figures ranged from Pecos County with 3,875 (3,000 of them at Fort Stockton on the opposite side of the county and far from the scene of the action) to Crane County with only 37 inhabitants. Both Reagan and Upton counties, the other two players in the oil activities, boasted of 357 and 253 citizens respectively.[16]

The housing that sprang up in those newly built oil boomtowns was of the same minimal quality as at all of the earlier booms. The shotgun layout was still utilized. Most of them were described as being of a simple framework construction with sheetrock interiors and a tar-paper exterior sheathing if they bothered to put on any exterior covering at all. There was also extensive use of corrugated metal for roofing, and most of the business structures were both roofed and sheathed in that ubiquitous metal material. Once again, living conditions were so difficult in the towns that most married contractors, like Joseph Graybeal, declined to move their families there. Instead, he recalled, "I never cashed a check. I sent all my checks to my wife and she would send me money. Whenever I wanted clothes, she knew my size—I'd tell her what I needed and she would ship them to me."[17]

The massive amount of activity caused the large oil companies to bring in a significant permanent labor force that they needed to house and feed. Their solution to the dilemma was to establish living quarters as close to the work sites as possible and away from the chaos of the boomtowns. William Allman pointed out the difference in the company workers' situation and that of the contractors. "The major companies then built camps and they kept their people out of living in those little towns. The contractors and the little people were the ones that had to live in those little towns." Although those clusters of housing were all termed *oil company camps* they ranged considerably in size from more than one hundred to as few as eight or ten houses. But all of them were self-contained in the sense of having their own water, electric power, maintenance facilities, and other amenities in order to maintain a comfortable lifestyle. Unlike company towns such as Texon they did not have a business district or a post office, although some of the larger ones maintained a commissary where foodstuffs and other supplies could be had. The large number and size of the West Texas oil company camps soon became a major characteristic of the region.[18]

Differentiating the population of those early Permian Basin boomtowns from that of the surrounding areas is extremely difficult due to the symbiotic relationship between them and the company camps. The population of the boomtowns waxed and waned according to the numbers of new discoveries in the area, which in turn affected the numbers of boomers who came and went. In the beginning almost everyone who came to those new towns expected to be there only a very short time. However, the numbers of camps surrounding the towns maintained a relatively stable population, which in turn provided an economic stimulus beyond that of the contribution of the boomers. For example, Jack Rainoseck recalled about Iraan:

> Most of the people who supported Red Barn and Iraan were living in camps. They had a small hotel, a small tourist court, and a few businesses. Of course the town would be up for six months and then down, but most of the people lived in camps. A small percent lived in town.

The same could be said for the area surrounding Crane, about which E. N. Beane declared:

> Yes, there were a number of camps. The Humble Pipeline Company had a big camp about couple of miles north of town and the Humble Refining Company, which is the producing department, had one up in the sand

> dunes two or three miles north of there. Cities Service had one; Magnolia had one; Tidewater Oil Company, they had one and they still have one out there. Then in 1929 Phillips Petroleum Company built their first gasoline plant in this area and they have maintained a camp there since 1929. I worked for them long enough to blow up the plant and I quit.

But the largest camp by far around Crane was the Gulf Oil Company's McElroy Camp that boasted sixty-five houses, six thirty-room bunkhouses, a hospital, and a clubhouse in addition to the usual complex of administrative, repair, and storage facilities. Taken as a whole the company camp system as implemented had a significant stabilizing effect on the boom activities in the region.[19]

At that time the primary activity spurring on the population growth was the massive amount of drilling taking place. All of the early drilling activities in the Permian Basin was of the cable tool variety and one of the necessary modifications to their activities was a gradual abandonment of the steam-powered rigs. The inability to find a reliable water supply for the boilers in that arid environment made it necessary to switch over to natural gas or diesel-powered engines and sometimes even to electric-powered engines. Wooden derricks also remained the norm for the first several years, but they also began to be replaced by steel drilling derricks as the decade of the 1930s approached.[20] Most of the drilling activity was in the 3,000- to 4,000-foot depth range, which utilized the standard cable tool drilling rigs. A notable exception to that rule was the Yates field in Pecos County where the wells were all in the 1,000- to 1,500-foot range. At that shallow depth it was more feasible to bring in portable cable tool drilling units like the National, the Star, and the Fort Worth Spudder to do the work, which in turn generally negated the need for rig builders in that particular area.[21]

Because drilling activity increased across the isolated region so rapidly and such a large quantity of oil was discovered the question of how to store and market the product became of immediate concern. It was necessary to build pipelines to transport the oil to market, but the far-flung and isolated production localities made that a slow and tedious process. In order to temporarily store the production a large number of 55,000-barrel riveted tanks were built at regularly spaced intervals throughout many of the fields near the producing wells to hold the production until pipelines could be laid. At least two of the one-million barrel concrete storage tanks fast becoming popular in the California oilfields were also built, one near McCamey and another at Monahans farther north, but they were never put into service due to cracks developing in their bottoms. All those large temporary storage tanks in addition to the massive number of smaller wooden and bolted steel production tanks brought large numbers of tank builders to the scene.[22]

As the pattern of newly established towns ringed by oil company camps, isolated drilling locations, and large storage facilities developed around those first

*Perhaps no other place better illustrates the difficulty of oil patch work in an arid and sandy environment than the area around Wink during the boom of the late 1920s. (Photo courtesy Southwest Collection/Special Collections Library/Texas Tech University, C. C. Rister Collection)*

discoveries, the drilling activity began to move farther to the north and west. In late 1926 a promoter named Roy Westbrook drilled the Hendrick field discovery well in Winkler County, which was at that time even more remote and sparsely populated than the fields already under production. By mid-1927 the Wink town site company began selling lots amid the sand hills in the Hendrick field and within a year more than two hundred rigs were running in the immediate area. Wink soon outstripped several other local town site promotion schemes and became the focus of the developing boom.[23]

The closest town with railroad connections to the newly discovered field was tiny little Pyote on the Texas and Pacific Railway more than twenty miles across the sand hills to the south. Overnight Pyote became a booming town estimated by some to have a population nearing 20,000, mostly transient citizens. Teamsters flooded in from the Cisco, Ranger, and Breckenridge area with thousands of horses and mules scattered all over the countryside in camps of more than one hundred animals each. They used twenty- and thirty-horse hitches to haul the heavy drilling equipment across the sandy land to the drill sites. To help solidify the unconsolidated roads they were graded and oiled in order to stabilize them for the heavy traffic. That unprecedented activity only lasted a couple of years before a railroad spur was laid from Monahans to the newly discovered field, and Pyote lapsed back into its preboom slumber.[24]

Meanwhile, drilling kept up a feverish pace and those first wooden cable tool rigs began to be replaced with steel derricks. C. L. McKinney noted: "When I first started shooting wells in Hog Town, Ranger, and Breckenridge and out at Colorado City and Crane County, they all had wooden derricks. That was along about '26. That's when they started using steel derricks." By late 1928 rotary drilling had also largely replaced cable tool activity. That early rotary drilling in the Permian Basin was conducted much like it was being done in other locations around the state. The wells were drilled down near the producing formation and then completed with cable tools for fear that the drilling mud would seal off the oil from the well bore.[25]

Also, by the time of the Hendrick development around the town of Wink the practice of shooting wells with nitroglycerin was also a routine practice in the West Texas oilfields. The "tight" formations of the newly opening Permian Basin fields, which did not allow oil to flow into the well bore in large enough amounts to make them particularly profitable, was well suited to the practice of well shooting. All the operators resorted to using explosives in order to fracture the rock of those tight formations, which allowed oil to flow into the wells in larger amounts. The nature of those West Texas wells required unusually large shots ranging from 800 to 2,000 quarts of nitroglycerin per well, dependent upon the nature of the formation being shot.[26]

The experienced hands who flocked to the new boom came primarily from the Ranger, Albany, Desdemona, and Breckenridge areas in north central Texas. By mid-1928 a significant number also arrived from Borger in the Panhandle as the activity slowed down there. The irresistible siren song of better wages and plenty of work drew them at a fantastic rate. All the publicity around the new riches being found in West Texas also brought its share of workers new to the oilfield as well as that usual unscrupulous element that always showed up around an oil boom.

Most consider Wink to be the wildest of the West Texas boomtowns, which was probably due to the rough element that was expelled from Borger transferring their well-organized criminal activities to the Winkler County town where they gained control early in its existence.[27] Ellis Summers, who served as a law enforcement officer there stated, "Everybody was used to a wide open operation. We tried to keep it semi-straight. The bootleggers and their sort were not paid much attention to at first."[28] A good example of the organized bootlegging situation can be found in the story that early day trucker Chris P. Fox related. It seems that all the freight that arrived at Pyote from El Paso was not what it seemed, as he discovered much to his dismay after some of his cargo was confiscated. He related:

> What they found out was that the bootleggers were making heavy boxes about 4 feet square and lining them with tin. Absolutely waterproof. In

> those days cases of Mexican whiskey bootlegged across the river at El Paso cost about a dollar a quart bottle. In the oilfield it brought about ten dollars. The box of liquor would be delivered to a loading dock at some mercantile store or pipe supply place in El Paso, billed to Pyote or Wink, and we would unwittingly load it on one of our trucks.[29]

However, in a little more than two years after the Wink boom began a combination of the falling price of oil, salt water encroaching on the newly drilled wells, and the gigantic boom developing in East Texas caused a general collapse of major oil development in the Permian Basin. As a result most of the newly arrived workers pulled up stakes and headed east for the new El Dorado in East Texas.[30]

The lull in West Texas oil drilling activity lasted until the mid-1930s. Beginning in 1933 and extending until the end of the decade oil discoveries in Ector, Winkler, Andrews, Gaines, and Yoakum counties in the western and northern portion of the Permian Basin created a rejuvenation of activity, and the workers began to return to the area. Odessa, the county seat of Ector County located between those new discoveries to the north and west and the original finds at Big Lake, McCamey, Iraan, and Crane to the south, began to develop as the major oilfield supply center for the entire Permian Basin. Meanwhile, Midland, some twenty miles to the east of Odessa, started its rise as the administrative center for the oil companies operating in the Permian Basin beginning with the 1932 move of Humble's personnel there from the McCamey area.[31]

Population trends in the region during the period 1920–40 clearly illustrate the changes underway. During those twenty years Midland, which started as the regional ranching financial center, grew from 1,795 to 9,352 residents, while Odessa, which began the period as a little ranch town of only 730 residents, developed into an oil worker center of 9,573. At the same time Midland County went from 2,442 residents to 11,721, but Ector County jumped from 760 to 15,051. During that same period those first oil discovery counties to the south, which were virtually deserted prior to the discovery of oil, stabilized as the boom passed. Crane went from 37 in 1920 to 2,221 in 1930 to 2,841 in 1940. Pecos went from 3,857 in 1920 to 7,812 in 1930, to 8,185 in 1940. Reagan went from 397 in 1920 to 3,028 in 1930 to 1,997 in 1940. Upton went from 253 in 1920 to 5,965 in 1930 to 4,397 in 1940. Meanwhile, those counties to the north and west boomed between 1920 and 1940 with Andrews going from 350 to 1,277, Gaines from 1,015 to 8,136, Winkler maintaining its earlier status at 6,141, and Yoakum growing from 504 to 5,354 with the bulk of the gain for all areas being between 1930 and 1940. The focus was clearly moving from south to north with Midland and Odessa emerging as the predominant oil towns of the region.[32]

By the time that drilling picked back up in the Permian Basin in the mid-

1930s transportation had completely changed the region due to improved roads and the introduction of more powerful trucks. The colorful sight of large camps of horses and mules and scores of heavy-duty freight wagons was a thing of the past. Instead, big heavy-duty trucks like Diamond T and Whites were doing all the hauling. Additionally, the introduction of caterpillar-type tractors that could grade down the sand hills and help provide all weather oiled and caliche lease roads in the midst of the newly opened fields gave ready access to heretofore hard to reach locations. As early as 1927 a useable road was graded the thirty miles south from Odessa to Crane, and prior to 1935 it was paved. That same route was extended farther south to McCamey during the same time period. The expansion of that network in other directions soon precipitated the concentration of trucking and earthmoving companies in Odessa.[33]

That renewed activity also sounded the death knell for large-scale cable tool drilling in the Permian Basin. With many wells going to depths of 8,000 feet or more it was almost impossible for cable tools to operate at that depth with any degree of efficiency. At the same time the introduction of the tri-cone rotary bit and other rotary innovations made that style of drilling much more efficient and it became the norm. With that change in drilling practices the cable tool driller and his toolie passed from the scene to be replaced by the roughneck who within a few short years assumed the legendary role of the oilfield worker best known to the general public.[34]

The 1930s was also a time when the traditional twelve-hour drilling tour was abandoned in favor of three eight-hour shifts, generally referred to by the hands as daylights (8:00 A.M. to 4:00 P.M.), evenings (4:00 P.M. to midnight), and graveyards (midnight to 8:00 A.M.). The daylight driller was considered to be the senior person on the job and when the crew was either "rigging down" or "rigging up" he was the man in charge of the combined crews doing the work. The changeover from a two-shift operation to a three-shift situation also brought the drilling crews into an hourly wage situation instead of a set amount of pay for a day's work. All that coincided with the Federal Fair Labor Standards Act of 1938 that established a minimum wage and set the number of hours worked for a week, after which overtime was paid.[35]

By 1935 safety concerns were also becoming more of an issue. "In traditional Texas oil-field culture, accidents were a part of the day's work and concern for safety was a sign of weakness."[36] There were even sayings in the patch like, "If you kill a mule you have to buy a new one but if you kill a man you just go out and hire another one," or, especially during the 1930s, "There is a man out there living on a soda cracker a day just waiting for your job."[37] With those kinds of ingrained attitudes it was hard to effect change, but beginning around 1930 major companies like Humble and Gulf introduced formal safety programs that trickled down to the contractors over the next few years. During that period moving parts on drilling machinery and oilfield equipment in general became covered with safety shields, which resulted in a significant reduction in oilfield accidents. It was also a time marked by the general use of steel safety hats by roughnecks. According to Arthur Stout, before then "we didn't have iron hats, just old felt hats, so we put a bunch of newspapers in the top of those hats."[38]

In the midst of this renewed activity there were efforts by the labor movement to organize the oilfield workers into a union. The only major effort by the union movement at organizing all the oilfield workers dated back to 1917 when a ninety-day strike waged by the AFL against the oil producers on the Gulf Coast proved unsuccessful. There were small sporadic efforts by the CIO to organize the West Texas workers early in the 1930s, with little or no effect. The one exception to that was the rig-building trade. As early as 1927 they were organized enough in the Panhandle to call a successful strike for higher wages. Accordingly, in mid-1936 the only major union effort in the Permian Basin developed when the United Rig Builders Union struck on June 10 for a general two dollar per day pay increase. The drilling contractors had recently raised drilling crews' wages by one dollar per day and the rig builders felt they were justified in their demands. The strike quickly spread across the Permian Basin and by the end of a few weeks brought drilling to almost a total halt. Although successful in securing a pay raise for the rig builders there are those who thought the incident hurried the development of unitized rigs, which ultimately resulted in the demise of rig building as a legitimate oilfield trade.[39]

Other changes began to affect oilfield workers' lifestyle during the depression years. One of the more significant was the introduction of trailer houses or mobile homes as living quarters. As early as 1936 the Stephens Hiway Home Company of Kansas City began advertising fourteen-, sixteen-, and eighteen-foot models that provided "real comfort at little cost for field workers." The use of those devices caught on by the end of the decade when the practice began to make a difference in solving the housing problems of those who followed the oil strikes across Texas. For the first time the workers had guaranteed living quarters, which tended to attract more stable married men with families to the oilfield work.[40]

It was also during that period when another innovation entered the industry that was destined to have a significant impact upon the colorful oilfield occupation of well shooting. Beginning in 1934 several oil operators in the Breckenridge area began injecting hydrochloric acid into wells that had gradually begun to decrease in production. In limestone and dolomite formations the acid penetrated crevices in the rock, allowed the oil to begin to flow, and rejuvenated the wells. When the Wasson field developed in Gaines County in 1937 those wells were usually not shot, but they were acidized on a regular basis with a considerable amount of success. Eddie Chiles, who was a pioneer in this aspect of treating oil wells, started his firm, which he named the Western Company, in the Wasson field and stated that by 1939 he was treating most of the wells completed there and in the surrounding areas. The process was less costly than fracturing the well with nitroglycerin shots, and it left the wells relatively undamaged for later maintenance work. This innovation marked the beginning of the decline of the work of the well shooter whose expertise was no longer needed.[41]

During that decade of the 1930s two more oil towns appeared in the Permian Basin. In 1935 Goldsmith was established in the oilfields about twenty miles northwest of Odessa when experienced town promoter Harry Tucker saw an opportunity to make a profit in the burgeoning oilfield where both Gulf and Phillips had established large camps. By the end of the decade it was a thriving community whose economy was based on serving the surrounding oil activity. Drilling was so active that the countryside was dotted with rigs. One worker described the nighttime scene as "Anyways you looked you just saw rig lights galore."[42]

A few years later, in 1939, Denver City became a reality on the southern edge of Yoakum County. It served the workers in the Wasson field, which was booming just a few miles away. Denver City was a true old-fashioned oil boomtown. According to Doc Cotton, "At first there were no houses, only a store and a service station." With housing in such short supply a large gas flare located at the north end of town provided a warm place to slumber as hundreds of men poured into the area. LaCosta Ivey described it as "a big flare down at the end of Main where men slept in cardboard boxes during cold weather—forty or fifty men at a time." Most who made the boom remember that when housing finally arrived the town became filled with tents, shacks of all descriptions, numerous cot houses, and a large number of the trailer houses that were becoming more and more common in the oilfield. The town site was ringed with at least eight oil company camps where the company men were housed while the contractors and their workers filled the new town to capacity, with many of them spilling over into Seagraves some ten miles to the east.[43]

Thus on the brink of World War II the Permian Basin was experiencing renewed prosperity and the future looked bright. What had begun with a rank wildcat venture in a remote area in 1923 had expanded across a wide expanse of West Texas, created several oil booms, saw the establishment of a half-dozen new towns, and populated a very sparsely settled region. Beyond that, those who were ranching in the Permian Basin region prior to the discovery of oil generally agree that the coming of the petroleum industry saved area ranches. Elliot Cowden said of the ranchers that "it saved many of them from being broke. They were so heavily in debt that I doubt they ever would have gotten out." Others, like George Bentley, went even further when he stated that "it was their salvation. They never could have survived without it."[44]

Beyond participating in the economic salvation of a significant portion of

West Texas the oilfield hands experienced considerable change in their lifestyle during the eighteen-year period from 1923 through 1940. During that time span the drilling industry changed over from predominantly cable tool to rotary drilling, steam power was largely abandoned, and the rigs went from wood to steel construction. Freighting and earthwork abandoned animal-drawn transport for cars, trucks, and self-propelled earthmoving equipment. The introduction of automobiles and the building of passable roads to accommodate them allowed workers to live farther from their workplace. The introduction of trailer houses began a change toward workers being able to follow the booms without being at the mercy of substandard local boomtown housing. Additionally, working conditions, including the length of the tours, how pay was figured, and the establishment of safety provisions, began to have an impact on the nature of work in the oil patch.

The groundwork laid during the late 1920s and the 1930s represented a change that would mark profound differences in the manner in which oilfield workers would live and work over the next couple of decades. The nature of the land in West Texas, with its great distances and arid environment, had much to do with those innovations, but profound technological change in the industry was most responsible. Through it all though, the hands remained the same rough, tough, and hard-to-bluff bunch they had been since oilfields first developed in the Lone Star State.

CHAPTER 8

# Change Comes to the Oil Patch, 1941–1960

As the decade of the 1940s began, the petroleum industry was still recovering from the twin effects of the Great Depression and the collapse of oil prices caused to a large extent by the immense production of the East Texas boom. Additionally, by 1941 the price of oil was still stagnant and drilling in West Texas was not escalating. However, the situation of the oilfield hands had some bright spots with plenty of work in the offing. Then, with the official outbreak of World War II in late 1941, the demand for oil skyrocketed, but the industry remained hamstrung by government price and production controls.[1]

However, getting oil out of the ground during the war years presented an entirely new set of problems for the oil patch. The war effort curtailed the acquisition of steel to the point that no new rigs were allowed to be built and drill pipe was almost impossible to obtain. Drawworks, pumps, and other types of drilling equipment had delivery dates going out at least a year, and drill pipe required a lead time of nine months to a year under the best of circumstances. Despite recycling outmoded equipment and patching up existing rigs and production equipment there was a significant drop in the number of operating rigs during the war years. On the one hand, rotaries were especially hard hit because of the large amount of steel required to supply them with the drill pipe necessary for them to drill. On the other hand, the cable tool units that were still being used in the older more shallow fields suffered less devastation due to not needing drill pipe to operate. In general drilling was restricted to proven areas and large-scale wildcatting was discouraged. The result was an ongoing battle between the oilmen and the government for allocation of materials that continued right up until the end of the war in August/September of 1945.[2]

With most of the able-bodied young men drafted into the army it became increasingly difficult to find workers to drill the wells and all the other jobs needed to operate the oilfields. The government did give some military deferments to experienced workers as essential personnel for the war effort, but those were far too few to fill the demand for oilfield labor. That situation was further exacerbated by the fact that despite being classified as an essential industry whose employees were eligible for draft deferments, young able-bodied oilfield workers were generally reluctant to get involved in the process due to the stigma of being looked upon as draft dodgers. Consequently, the oil patch was forced to continue the rapid pace of the work with inexperienced older men and teenagers supplying much of the manpower. Workers ranging from truck drivers to rig builders and everybody in between were in short supply. By 1944, when

3,000 rigs were needed to drill the number of wells forecasted by the government agency PAW (Petroleum Administration for War), and with the numbers of operable rigs dwindling, the pleas for additional men and supplies became more and more strident from industry leaders. Oil-related industries were urged to train men over the age of thirty-eight and under twenty for various jobs. Women were recruited by some of the larger companies to work at jobs deemed suitable for their physical limitations, and in general every available manpower resource was explored. Workers were assured that the regulations of the relatively new Fair Labor Standards Act would cover practically all oilfield workers under the rule of minimum wage and time and a half pay for hours worked in excess of forty each week. During those years most of the petroleum industry employers worked their employees from forty-eight to sixty hours per week.[3]

The situation of the drilling contractors was particularly difficult due to the manpower dilemma. Their practices are exemplified by the situation that developed during 1944 in the newly opened TXL field about twenty miles west of Odessa. Drilling contractors descended upon the area like a plague of locusts and rig lights lit up the area every night.[4] J. D. Brown, an experienced roughneck, remembered working derricks at TXL during the height of the drilling and making a deal with the drilling company to work on five rigs at the same time. This is how he explained it:

> I was working derricks out there at TXL during the war and we was always short handed. Everybody out there worked all the doubles they could stand and a lot of the time we even drilled with a four man crew. Other than the drillers and three or four of us other hands they was mostly weevils and nobody wanted to work derricks. Then one day the tool pusher come to me and made me a proposition. He allowed that if I stayed out on the job twenty-four hours a day all I would have to do was work derricks every time they needed to trip on one of his five rigs. The rest of the time I could sleep in the dog house and he would pay me for all the time I was out there. Well sir, I spent three weeks out there sleeping in that damned dog house and living on baloney sandwiches. I thought I would never see another poor day.[5]

It was definitely a time when a roughneck could make some good money if he could stand up to the work. During that same period Ralph Thompson attended high school in Odessa and roughnecked evening tour on rigs operating close to town for at least a year. He was a big strong kid for his age and he remembered:

> They picked me up every day outside the school house. When we finished tour they would drop me off at my house just after midnight. By the time

> school was out that year the driller even had me working derricks part of the time and I had more money that I had ever seen before in my life.[6]

Although stories like these are common among the workers it must be remembered that while drilling contractors were willing to hire just about anybody who could walk and talk, the major oil companies declined to hire underage workers and were much more circumspect about their general hiring practices.

A similar situation existed up at Denver City, which had boomed back in 1939 and was still very busy. Although there were a number of oil company rigs running, the contractors were there in full force. One eyewitness stated: "You could count as many as one hundred twenty five rigs running from any one spot in the field." With that much drilling activity going on Buff Ivey, who was born in the area, recalled:

> In 1943 most of the teenagers from around here roughnecked. They was paying good wages and they was plenty of work and you could double anytime you wanted to. Back then if you needed a roughneck you would go to the pool hall and find one. Once I just took a few days off and when I was ready to go back to work I went to the pool hall and I seen this guy I knew. He was a tool pusher over seven rigs. He said "I can give you any rig you want and any tour you want. Just whatever you want." It was that easy to get a job.[7]

"Doc" Cotton, who was working for Humble in the Denver City area at that same time, elaborated on the drilling situation. He observed that although many of the roughnecks lived in Denver City just as many or more drove the thirty-five miles from Hobbs, New Mexico, or the fifteen miles from Seagraves, Texas, to their work sites on a daily basis over roads that were little more than sand beds. He allowed that during the war years, "We had to make do with what we had" in both equipment and men. Despite the severe manpower shortage Cotton stated that although "contractors used a lot of young boys, the oil company's didn't. I don't remember a time that Humble hired any of those kids."[8]

Despite the fact that paved main highways and better-quality automobiles allowed roughnecks to live farther from their rigs and drive to the locations, oilfield roads remained rough and unimproved. Pete Sitton commented on that when he stated, "Roughnecks had to have a car. Used to be if your rig moved your hands moved with it. In 1942 I moved seven times in one year. Back in the days before the war in a radius of three or four miles might be thirty or forty rigs running," but all that began to change in the late 1930s and early 1940s.[9]

While the roughnecks enjoyed all the success they could say grace over during the war years the pipeliners did not fare as well. The severe shortage of steel

*Night scene at TXL when the boom was on during World War II, showing the proliferation of rig lights that fascinated so many. (Photo courtesy Southwest Collection/Special Collections Library/Texas Tech University, C. C. Rister Collection)*

greatly restricted oil company pipeline construction and most that did take place utilized used pipe. A good example of the curtailment of that work is that of the Humble Pipeline Division, which laid only 150 miles of pipeline between 1942 and 1945. The major exception to that situation was the laying of the various government-sponsored lines. The first and most publicized of those was the "big inch." It consisted of a twenty-four-inch pipeline financed by the government as a war emergency project. It was designed to run from near Longview, Texas, to the New York refining areas, and its first section that ended at Norris City, Illinois, was completed late in 1942. That first section of "big inch" that was completed between June and December of 1942 exemplified the speed and efficiency with which the newer mechanized ditching and laying equipment combined with the latest welding techniques could lay pipelines. They averaged nine miles per day on that job. It also used significantly fewer workers per mile of laid pipe than only a few years earlier. In all, the PAW authorized thirty-three of those types of projects for the war effort.[10]

Tank builders, much like the pipeliners, did not achieve a spectacular increase in their work during the war years. According to "Whitey" Harding who had built tanks in the Permian Basin since the early days at Best and Santa Rita: "We had plenty of work in those days, but it was mostly secondhand stuff. We cut down a lot of old tanks and reset them on account of the steel shortage. Besides, most of the help was either weevils or winos. All the best hands went to war." Another tank builder, Bill Walker, managed to fare a lot better in the 1940s when he was sent to North Africa to build "a whole raft of ten thousands" for the military. He had to use native labor for the work and he swore that those hands "were so slow that the damned things like to have rusted down before we finished them."[11]

Rig builders were almost as busy as roughnecks during the war years despite many of the drilling rigs being skidded from one location to another in the mostly level country of West Texas. The process had become a lot easier since those days of using mats or lay boards and rollers during the early 1930s in East Texas. In 1942 Gene Rumbaugh, who was operating out of Odessa, introduced a skidding system for rigs that consisted of placing a large dolly similar in construction to the treads on bulldozers under each derrick leg and pulling the mounted derrick with a bulldozer. His firm even cut down sand hills with dozers in dune areas and skidded over them with considerable success. Of course there was the occasional hitch, as witnessed by W. R. Johnson, when one of the dollies became dislodged while skidding a 120-foot derrick in the Goldsmith area. He said it "swayed about three times, leant to the north, and fell. It sure made a terrible mess." Despite skidding being faster and cheaper, uneven terrain and

*Side booms lowering welded pipe into a ditch during the big inch project. Mechanical innovations like the side boom hoisting and lowering apparatus both decreased pipeline crew size and increased the speed of the work. (Photo courtesy Southwest Collection/Special Collections Library/Texas Tech University, C. C. Rister Collection)*

long distances dictated that the bulk of the work of moving rigs be done by the rig builders who dismantled the derricks and rebuilt them at the new locations. The rig builders, about the only union organized group in the patch, even called a strike early in September of 1942 for higher wages.[12]

With all that activity going on both Midland and Odessa jumped in population from around 9,000 each to more than 12,000 as they solidified their positions as regional hubs for the oil industry. The rest of the Permian Basin towns maintained a relatively stable population during that period.[13] As people flooded into Odessa housing became extremely scarce, and the thing that seemed to impress everyone most was the nighttime scene. Carl Weaver commented on it when he said:

> Back during the war years you could go to a football game in Odessa and from up in the stands you could see rig lights in a circle all around the town. Those rig lights and the gas flares, they just lit up the countryside.[14]

Once again housing became a significant problem in the oil patch as a boom of a slightly different complexion developed, but this time the situation was handled somewhat differently. This time the government stepped in and attempted to alleviate the housing situation in light of the war emergency. For example, in Andrews, about thirty miles north of Odessa, the federal government established a two-city-block section called Deep Pay Village. It consisted of trailer houses with the wheels removed and placed on concrete pads in order to give them a permanent look. The facility was capable of housing two hundred families and filled up almost immediately. About the same time Victory Village was established in Odessa. It consisted of one hundred prefabricated permanent homes and it was filled as soon as it was completed. Both facilities were government-subsidized rental properties established specifically for oilfield workers in the area, and it was considered of the greatest urgency to make them exclusively available for oil workers in the interest of the war effort.[15]

Those projects were indicative of the general housing shortage across the nation that prompted the government to introduce a subsidized program for the use of trailer houses as temporary living quarters associated with critical war industries. It was particularly well suited to the oilfield workers who had begun to use those portable residences as early as the mid-1930s. Deep Pay Village in Andrews was an example of that policy, which was evidently very successful. This was also a time when the drilling companies began to use trailers for "dog houses" and other accommodations associated with their rigs operating across the breadth of the Permian Basin. It was also a time when the first trailer parks appeared in oil towns throughout Texas as a preamble of what was to come in the postwar years.[16]

During the course of the war PAW estimated that 75,000 new oil wells were needed to keep pace with demand. By the time the war ended in August of 1945 the petroleum industry had exceeded that goal and completed 85,000 of them despite labor shortages and deteriorating equipment. By then most of the oilfield equipment and especially the drilling rigs were reduced to little more than junk. It was a perfect time for change in the oil patch and change came quickly. The immediate postwar demand for more oil combined with significant technological advances within the industry created a situation that ultimately called for a better-educated worker to do the job. Between 1945 and 1955 practically all drilling rigs were built with internal combustion engines as a power source, and innovations on the drawing board prior to the beginning of the war became an almost instant reality.[17]

Those changes increased the speed and efficiency of the drilling process partly through the unitization of equipment and the development of telescoping masts on drilling rigs. The combination of those elements into what the workers referred to as "jackknife" rigs, because the derrick could be folded down like a knife blade and moved by truck, spelled the beginning of the end for the rig-building trade. Whereas it took a week or so to dismantle and rebuild the old standard rigs, the newer units could "rig down," move, and "rig up" in less than half that time without the use of the specialized labor of the rig builders. After 1945 no more standard rigs were built, and by the 1960s rig builders were gone from the scene as one of the first casualties of post–World War II technological change in the oil patch.[18]

Just as the rigs became more mobile with the newest innovations the drilling crews became more mobile due to improved roads and automobiles. During the war the practice of driving thirty or forty miles to work on a rig became commonplace, and by the beginning of the 1950s the activity expanded to become the norm. The bulk of the drilling crews in the Permian Basin headquartered in Odessa and drove to their jobs. It was not unusual for them to drive almost a hundred miles in any direction to go to work, but the average seems to have been between fifty and sixty miles with the occasional easy drive of only ten or fifteen. It is generally agreed among the hands that a hundred miles was about the maximum drive.[19]

Drilling contractors traditionally paid only for the time spent working on the rig. Consequently, in order to keep their expenses as low as possible the crews took turns driving to the job in what in an urban setting would be called car-pooling. Thus, a roughneck would only have to drive every fifth day. In actual practice some hands were many times temporarily without transportation or simply preferred not to drive. In those instances the driller might, as in the case of Pete Sitton, drive all the time and "charge the hands fifty cents per day if they didn't want to drive." By the mid-1950s most drilling contractors began paying

*The "jackknife" drilling rig, whose introduction sounded the death knell for rig builders and created the need for a better technologically educated drilling crew, is shown here, ca. 1948. (Photo courtesy Oklahoma Historical Society, Ray Jacoby Collection)*

their crews for driving time at a lesser rate, usually the minimum federal wage, than their regular pay. By around 1960 or so it became a general practice for many contractors to pay the driller for driving in order to relieve the roughnecks of the chore.[20]

Roughnecks worked seven days a week and received time off only a day or so now and then when they finished a well and moved to another location or when the rig was "stacked" for some other reason. All that driving took its toll on automobiles. According to Bob Cullen:

> Tires would only hold up for about 10,000 miles or so. They would generally last maybe six to eight months before they were ruined. We took pretty good care of our cars, but usually you could count on having to get a new one about every two years. Those old lease roads were so rough. Several times I ruined a tire the very day I bought it.[21]

During the 1950s hiring out as a roughneck remained in the traditional mode. If a hand arrived in town looking for work he generally headed for the pool hall where there was usually a chalkboard on the wall where drillers listed job openings along with their telephone numbers. Sometimes they were even more direct. Jerry Holt remembered a late evening pool hall event when a driller jumped up on a pool table and announcing in a loud voice, "I need a derrick man to work graveyards and I need him right now." He had one before he could get down off the table. In addition to the pool halls there were always cafés that served as labor clearinghouses. Those establishments each catered to a specific clientele. For example, in Odessa during the 1950s Dude's on East Second Street headquartered for drilling crews, and Janey's, also on East Second, attracted the tank builders, while Cookie's on West Second Street was where the roustabouts hung out. The rig builders congregated at the Carol Café on North Grant Street during those days. Most times those eating establishments put workers together with potential employers and provided meal tickets to the hands until they received their first paycheck. That check would be delivered to the café owner who received his money first and gave the rest to the worker. Sometimes the café owner or a fellow worker would simply buy the check from the worker ahead of time at a greatly discounted rate and have the check endorsed over to him. Either way the deal was consummated it was a system that dated back at least to the 1920s and worked well for those following the booms.[22]

The first major oil boom following World War II that approached the magnitude of those of the 1920s started when a big producer was drilled on the Canyon Reef Trend in Scurry County of West Texas in late 1948. By the following spring the closest town, Snyder, was in the throes of a major oil boom. Within the next twelve months there were 187 rigs running in the general area, which translates into almost 3,000 roughnecks, not to mention all the other oilfield trades, working around the clock.[23]

Although thousands of oilfield hands rushed to the site and the town exploded in population from 7,000 to 20,000 within a year, the Snyder boom never developed the rough reputation of earlier booms like Burkburnett, Ranger, Borger, and Wink. There were some sound reasons for the more controlled situation at Snyder that help explain the changed nature of oil booms. First of all Snyder, in keeping with the pattern of some of the previous oil booms, was already an established community with a strong infrastructure in place that was characterized by a robust law enforcement element. Also by that time the petroleum industry had attracted a much larger complement of stable married workers, many of them recently returned war veterans, who followed the work in the developing oilfields in their mobile homes. Those who made the Snyder boom reported seeing thousands of trailer houses lined up throughout the town, and there was an almost total absence of the shacks and tents that had composed the primary housing in prewar boomtowns.[24]

Samuel D. Myers commented on the importance of mobile homes as a stabilizing factor during the Snyder boom.

> The most important living quarters were trailers, which by now had come into general use. It is estimated that 3,200 trailers were stationed on the streets, on vacant lots, and in parks provided for them, during the most active months in the oilfield nearby. The mobile home units are generally credited with preventing a housing crisis of serious dimension.[25]

In addition to the thousands who flocked to Snyder to work, many others opted to live either at Colorado City or Sweetwater and drive the thirty to fifty miles to their work locations. Some even made the daily commute from as far away as Odessa, which was approximately one hundred miles distant. Others, like Matt Ware, who was building tanks for the National Tank Company at the time, stayed in motel rooms in Snyder that the company rented for a year in advance to assure a place for the hands to stay.[26]

Many of the tank builders had permanent homes in Odessa, and with lodging so difficult to find in Snyder drove the hundred miles regularly. By leaving home around 5:00 A.M. they could be on the job by eight and be back home before it got too dark. Work was so plentiful that they remembered driving past six or seven tank locations with the steel stacked on the ground ready for assembly before arriving at their work site. With that much work available and the need to work seven days a week a workingman could do very well. In an era when normal conditions provided an income of $6,000 to $7,000 per year for tank builders they were making in the $12,000 to $15,000 range during the boom.[27]

Hard on the heels of the Snyder boom, which tapered off early in 1951, came another boom generally referred to as the Spraberry boom. As it developed across a seven-county region with the greatest production located to the south of Midland, more and more workers were attracted to the Permian Basin. The Spraberry Trend, as the production area was known geologically, began being developed during the summer of 1951, and by that winter work was in full swing. It was destined to remain active until around 1954.[28]

Working conditions at Snyder and particularly during the Spraberry boom were indicative of the changes in the way that labor would be conducted henceforth in the massive West Texas oilfields. The Spraberry finds were located in sparsely settled ranching country closely akin to those original 1923 discoveries in the Big Lake/Texon area. The closest urban centers of any size were Midland and Odessa, which were fifty to eighty miles distant. Despite the absence of any towns close to the activity the general improvement in transportation capabilities mitigated against town site development in the midst of the drilling activity, as had been the case throughout most of oilfield history in the state. The bulk of the workers lived in Odessa, which had already achieved the reputation of a workingman's town although a considerable number did headquarter in Midland.

As work progressed during the Spraberry boom the dry alkali soil of the country was soon reduced to a fine powder by oilfield traffic. All who made the boom remembered choking clouds of dust raised by the thousands of cars and trucks traveling across the arid ranch land. As those simple graded roads became rutted and littered with potholes the oilfield drivers simply spread their driving out to avoid the rough terrain and clouds of dust. In places those roads became a hundred yards or more across.[29]

The devastation created by all that activity caused more than one incident between the oil people and the local ranchers. Matt Ware recalled one that occurred when he and his tank-building crew were working in the Driver Field.

> We finished work this particular day and started to drive out to the main road. When we got to the cattle guard this old rancher was standing there with a .30–30 Winchester cradled under his right arm. He flagged us down and said, "Boys I know you are just a bunch of working men and I hate to do this to you, but if you want back in this pasture tomorrow it will cost you five dollars per car and twenty dollars per truck to get in." Well sir, I asked him if it would be all right for us to go back over to the location and load our tools and get the compressor. He said that would be fine and when I got back home I called the branch manager and told him what had happened. Evidently the various companies and the rancher got together that night and settled their differences because we went back the next day and kept on working. Of course after that we were a lot more careful about keeping to the main road.[30]

Perhaps the single most important thing that shows the change in the work patterns engendered by transportation progress are Permian Basin population figures for the fifteen years following World War II. At the end of the war both Midland and Odessa had grown to 12,000 citizens each while all the older oil towns maintained a relatively stable population that was considerably smaller than their boom days. As the 1950s began, of those entities outside Midland

and Odessa only Andrews County showed significant growth with a population that went from 1,277 to 5,002 between 1940 and 1950 and that more than doubled to 13,000 during the decade of the 1950s. Snyder boomed from its original size of 7,000 to over 20,000 by mid-decade and remained relatively stable at that number until 1960. Only Midland and Odessa really grew during the 1950s. By mid-decade, at the end of the Spraberry boom, they were at 35,000 and 45,000 respectively. By 1960 every area in the Permian Basin was in a holding pattern as far as population growth was concerned except Midland, which reached 62,000 as the regional headquarters and administrative center for oil companies, and Odessa, which had soared to 80,000 as the operational base for labor across the entire Permian Basin.[31]

The explosion in population thrust both those major Permian Basin urban centers into severe housing shortages similar to but not as extreme as the boom situations of oil towns of the past. A proliferation of shacks and other temporary shelters never developed in those towns. Instead major home-building programs were instituted in both Midland and Odessa that alleviated the situation somewhat but did not completely solve the problem. The overflow population began to gravitate toward living in trailer houses very similar to the situation that developed during the Snyder boom. Those mobile homes that were particularly favored by drilling crews whose seminomadic way of life lent itself to making trailers their permanent residences. In evidence of that trend a survey of trailer parks in Odessa, which served as the hub of oilfield work in the Permian Basin between 1946 and 1960, indicates that there were only two trailer parks in the city in 1946. Within two years that number had risen to thirty-four and by 1960 the town boasted sixty trailer parks that housed an estimated population of 10,000 in a town of 80,000 total residents.[32]

That increased mobility also had other unexpected impacts on the workers in the region. One of the major characteristics of Permian Basin oilfield life for decades had been the existence of oil company camps that had developed a culture of their own. There were hundreds of those camps scattered across the oil patch of the Permian Basin. Beginning in the 1950s and continuing until around 1960 all those facilities, some of them in existence for as much as thirty years, were dismantled. It became more economical for the oil companies to end the considerable expense of maintaining the properties when their employees could easily drive to their work places at little or no expense to the company. In most cases the employees were allowed to purchase the home in which they had resided for years at some minimal price or at no cost at all beyond the expense of moving it to a new location. It was a devastating blow to most of the workers who had lived in the camps for years, raised families there, and never dreamed of living anyplace else. Additionally, they were accustomed to paying some pittance, usually in the twenty dollar per month range, for rent,

*Odessa, ca. 1920, before the discovery of oil when it was a little cow town of approximately seven hundred residents and the edge of town was a stone's throw away from the courthouse. (Photo courtesy Southwest Collection/Special Collections Library/Texas Tech University, Jack Nolan Collection)*

*Odessa in the mid-1940s when it was in the 12,000 to 15,000 population range and had assumed the role of oilfield labor and supply center for the Permian Basin. (Photo courtesy Southwest Collection/Special Collections Library/Texas Tech University, Jack Nolan Collection)*

utilities, and upkeep. It caused them considerable discomfort when they were forced out of their comfortable nests. As they bought lots in nearby towns and prepared to move their houses many of the dispossessed workers developed a great deal of animosity toward the towns, which they considered were charging them an unreasonable fee for taxes, utilities, and all the amenities that they had been getting for free for so many years. It was a traumatic time; for many of the oil company people it left significant psychological scars that took years to heal.[33]

In the excitement of Snyder and Spraberry most of the hands didn't pay much attention to the many technical introductions that began to change their way of life. They were probably too tired after putting in those long days to care about anything beyond getting a little rest. But during the latter half of the 1950s those changes began to have an impact. For example, the tank-building trade slowed when welded steel tanks in the 200- and 1,000-barrel range for use as field tanks were introduced to replace bolted tanks of the same capacity. Then, late in the decade, fiberglass tanks appeared to replace the wood tanks that had traditionally served to contain the more highly corrosive oilfield products.

Welded tanks were constructed in a shop and hauled by truck to the sites where they were set and leveled and walkway and stairway installed. It was the type of work that could be performed by roustabouts and other unskilled labor. Jimmy Zeigler went on at length about those changes as early as 1958 when he reminisced about tank building.

> I started tanking right after the war about the time we stopped using ratchets and went to compressors and impact wrenches. Back then I worked in a steel crew. I never did work in a wood crew. Back then we built mostly new stuff and occasionally cut down and reset a few of those old greasy secondhand tanks. Up until just a few years ago you could make some good money tanking cause we contracted all those bolted jobs. Now about half the work is setting those damned welded tanks and building a little walk and stair. You see contracting a man can clear about $25.00 a day, but all that welded stuff pays day labor and a man is lucky if he can make as much as $15.00 for a full days work. To make matters worse when you build a couple of bolted hi-fives it takes two days and you can set two of those welded five hundreds in a single day. Lately I've been lucky to go out on a job more than a couple of weeks a month.[34]

"Peewee" Johnson, who had set tanks for years, elaborated on the change in the work about the time he stopped tanking full time and started a small roustabout contracting company in Odessa.

> Work got real slow about '53 or '54 and I got a chance to contract some grain tanks at elevators up in Oklahoma, Kansas, and Colorado. For about three or four summers I took as many as ten or twelve tank hands up there and we built a bunch of eight, ten, and twelve ring bolted tanks. By the time that played out work here in the patch was still slow so I started this little outfit here. We set a few tanks and do some repair work, but mostly we handle regular roustabout connection work. There just ain't enough tank work to keep busy.[35]

As the decade of the 1950s ended it was obvious that tank building, which had been a viable trade in the oil patch since the industry's beginnings in Texas, was fast disappearing.[36]

Another of those traditional oilfield trades that saw its demise in the 1950s was that of well shooting. Nitroglycerin shooting of oil wells to stimulate production began in the nineteenth century with those first wells in Pennsylvania and was the only method used to stimulate wells until the introduction of the acidizing process at Breckenridge, Texas, in the 1930s. Shooting was one of those jobs considered to be so dangerous that very few workers became involved in it, but with the changeover from liquid to solid or jellied nitroglycerin at the end of World War II it became much safer. Gradually, as newer techniques like acidizing and sand fracturing became more and more efficient shooting faded from the scene.

C. L. McKinney, who began his shooting career around 1920 in the Ranger, Desdemona, Breckenridge area, was one of the last shooters operating in the Permian Basin. He gave a clear description of the end of an era when he stated:

> You see I didn't go into business for myself until 1945 and what helped me a lot was they were already making good built-in shooting equipment. I had a pretty good business around here in West Texas and in 1950 I went out to New Mexico where the big boom was in the gas field. When it shut plumb off around here it was about '54 I believe. That's when I went out of business was in '54.[37]

Those twenty years from around the beginning of World War II until 1960 marked the end of many of the ways things were done in the Texas oil patch since its beginnings at the turn of the twentieth century, due in large part to advancement in transportation. Making travel easier and the development of mobile homes played a huge role in ending the traditional image of oil boomtowns. That same transportation revolution directly resulted in the demise of the oil company camp system that had played such an important role in the life of oilfield workers for decades. Also, general technological progress of the era caused the demise of at least three traditional oilfield trades: rig builders, tank builders, and oil well shooters. It was a time when life became easier for the hands in the oil patch, who for the first time were able to settle down to a more stable lifestyle and enjoy the reputation they had created as oilfield trash.

CHAPTER 9

# The Making of an Oilfield Culture, 1901–1960

At the dawn of the twentieth century Texas was poised to enter an era of unprecedented economic prosperity dominated by the petroleum industry. At that time 82 percent of the state's 3,048,710 residents were classified as rural, and the only oil production was a small operation clustered around the town of Corsicana in Navarro County. The expanding petroleum industry in the United States, dating back to the 1859 Drake well in Pennsylvania, was barely forty years old. Its substantial, although not giant, production supplied a growing number of refineries whose output consisted primarily of kerosene for lighting, lubricants, and some gasoline for the internal combustion engines that were gaining in popularity. Wooden drilling derricks had been standardized in size since the 1880s and rotary drilling techniques had been used in oil well drilling for less than ten years. The drilling equipment of the day, although suitable for drilling below 2,000 feet, seldom completed wells deeper than 1,000 to 1,500 feet. All this was about to change dramatically.

In January of 1901, when the Lucas well blew in at almost 100,000 barrels per day and succeeding completions fell into the same range, a seemingly unending supply of oil became available. It was an unprecedented phenomenon that heralded a major change in future energy usage. Almost immediately petroleum industry leaders began promoting uses for their product, exemplified by the soaring demand for gasoline over the next ten or fifteen years as automobiles came on the market. The petroleum industry was poised for massive growth.[1]

The first twenty years of the twentieth century witnessed an oilfield culture develop in Texas that utilized a labor force largely made up of the existing rural population of the state. In the years following World War I that same group of oilfield workers solidified their position within the social fabric of the state. Amid the background of one oil boom after another that repeated the same monotonous story of immediate wealth, uncontrolled production, and temporary social dislocation, the industry grew and changed and along with it the workforce also grew and changed. Although all those booms superficially appeared to be the same, each of them had its own unique characteristics over time and through space. Over time the professionalism of the workforce grew and through space geographic characteristics demanded different approaches to the work. The combination of those elements along with technological change resulted in a workforce that tended to retain the rural attitudes of its beginnings as it progressed into a more highly specialized technological future.

One of the first characteristics of oilfield labor that developed was the division between the company workers and the contract workers. At the very beginning several oil companies were formed that eventually became huge entities. Gulf Oil Corporation (1901), the Texas Company or Texaco (1902), Humble Oil Company (1911), and Magnolia Petroleum Company (1911) were among those spawned in Texas, while out-of-state firms like Sun Oil Company entered as quickly as possible. These were the type of firms that came to be called the majors. Additionally, a host of other smaller companies of varying size were formed. They came to be called the independents. All of those companies, both majors and independents, bought leases in the newly discovered fields, produced the oil, and sold it to the refineries that were typically owned by the majors. All those companies, large and small, needed a stable and reliable workforce to accomplish their corporate goals. Hence, those workers employed by the oil companies enjoyed a more or less stable lifestyle and an assured paycheck based on an hourly pay scale, which attracted workers who tended toward being more family oriented.

Most of the work of drilling oil wells, laying pipelines, building tanks and rigs, and all the other types of labor relating to getting an oilfield ready to produce lay primarily with contractors hired by the oil companies. The major exception to that was that beginning in the 1920s and continuing until the 1960s most majors formed drilling divisions that they used to drill their own wells in order to meet lease obligations when boom conditions made it impossible to get contractors. The same could be said of roustabout work and well servicing.

Once a new discovery was made the various contractors flooded in and began the work of creating an oilfield. That massive introduction of contract workers into a newly opening oilfield created an oil boom where those workers were usually referred to as boomers by the company men and outsiders. If they wanted to work, those hands had to follow the booms even though none of them intended to remain beyond the duration of the boom. As soon as the initial large-scale drilling ended most of them left for another field opening somewhere else and the boom ended. This type of labor tended to attract young single men who could bear up under grueling labor conditions that, in turn, allowed them to enjoy much higher wages than was otherwise available to regional workers. Those higher wages were many times based on a daily or piecework rate of pay. The rough and ready reputation of those young oilfield workers caused them to be referred to as oilfield trash by many. It is that aspect of oilfield society as it developed in Texas that has colored the public perception of who those workers are and what they are like.[2]

Although the company employees were regularly transferred from place to place, just as the boomers moved from boom to boom, the most obvious difference between the boomers and the company men manifested itself in housing. In the first days of the Spindletop boom, workers made a clear distinction between the housing accommodations provided by the operators in the area. They much preferred working for those who provided clean dry sleeping and dining facilities to those who left the hands to their own devices in finding living quarters and eating establishments. That fact was not lost on the oil companies as they established themselves in those early days.[3]

By 1911 the finds around Electra drew oil activity away from the Gulf Coast to the north and west that culminated in the gigantic discoveries at Ranger in 1917 and at Burkburnett the following year. As this was developing the large oil companies began expanding their refining facilities on the Gulf Coast, which after 1918 became massive. At the same time they were trying to maintain a significant presence in the newly developing areas. The added pressure of manpower losses due to World War I during 1917 and 1918 made it imperative that the oil companies entice workers; good housing was one of the incentives to retain a viable work force. Additionally, the only major labor strike by Texas oilfield workers, which took place in late 1917 and early 1918 in the Beaumont area on the Gulf Coast, added emphasis to the need to attract and keep good employees. Although pay raises offered by Humble and others are largely credited with thwarting the goals of the unionized workers, a major factor in their lack of success was losing the support from refinery employees who were a major beneficiary of company housing.[4]

It was during this period that the oil companies began making significant efforts toward solving the housing dilemma that went along with the oil booms. In February of 1919 the Texas, Gulf Coast, and Louisiana Oil and Gas Association appointed a committee to inspect oilfields and report on living conditions. The committee members were appalled at the situation they found and reported back that the companies definitely needed to take steps toward providing better housing for their employees.[5] Meanwhile, the Prairie Oil and Gas Company had already built the first big oil company camp at Ranger early in 1918, but significant further construction was stymied when a few months later the government put strict controls on construction materials due to the war.[6] As the war ran its course the controls were loosened, and housing construction by the oil companies exploded along the Gulf Coast when the Sinclair Refining Company built a small town on the Houston Ship Channel complete with a school, a post office, parks, and other amenities. Meanwhile, both Gulf and Humble completed major housing projects near their growing refinery complex at Goose Creek.[7]

The opportunities for providing oilfield housing prompted the development of a number of companies specializing in supplying those dwellings. The E. L. Crain Lumber and Manufacturing Company of Houston was one of the more prominent of those companies. Beginning in 1918 they advertised sectional and ready-cut homes available to be erected in less than one-half the usual time at a 25 percent to 50 percent cost savings, and it was 95 percent salvageable if moved to another location. They had scores of floor plans that could be assembled by inexperienced carpenters. The Crain Company sold thousands of homes to oil companies on the Gulf Coast as well as in the opening North Texas fields and overseas operations. In 1923 they changed the name of their company to the Crain Ready-Cut House Company and remained active in oilfield sales until at least 1925. In addition to the Crain Company many others entered the market to supply the growing demand. The Texas Concrete Construction Company developed the Moore precast reinforced concrete homes that they supplied to Humble at their Baytown refining facility for company housing. Others like the Aladdin Pre-Fab Homes Company and PDQ Houses of Houston offered services similar to that of the Crain Company. By 1929 the Southern Mill and Manufacturing Company of Tulsa, Oklahoma, was also offering a wide range of their "Sturdybilt" sectional homes to all the developing oilfields in the midcontinent region.[8]

Despite the proliferation of companies offering prefab homes the greatest volume of housing provided by the large oil firms was built by company crews utilizing a standard house plan with a standard color scheme. An excellent example of one of those early camps was the one built by the pipeline division of Humble near Cisco, Texas, around 1920. Known as Humbletown, it consisted of sixty-five houses for families, five large bunkhouses for single men, a mess hall, a commissary, tennis courts, and its own gas, electric, and water facilities. All the structures were painted in the company colors of battleship gray and red. Family housing cost employees five dollars per room per month, which amounted to thirty dollars for two-bedroom houses and twenty-five dollars for one-bedroom accommodations. The bunkhouses and mess hall served 160 single men working on the pipeline; they paid one dollar per day for their room and board. Having access to those kinds of accommodations, which were as good or better than similar ones in settled towns, was indeed a luxury compared to the crowded and usually squalid conditions available in the boomtowns.[9]

In the early to mid-1920s, as oil activity spread into the Panhandle and West Texas, the sparsely settled nature of the country made company camps a necessity for the large oil companies. The problem there was not so much the necessity for decent housing under boomtown conditions as the fact that there were no towns at all near the scene of the discoveries. Those company camps provided little oases of shelter in a most dry and inhospitable environment, and they became a major element in the oilfield culture of the region. Those homes provided a sharp contrast to the tents and other temporary shelters used by the contract workers as they camped near their work sites or even the shotgun houses, shacks, tents, and other inadequate types of shelter afforded by the boomtowns to the bulk of the contractors in that region.[10]

By that time the large companies were also providing benefits other than

*This typical oil company camp in the midst of the Permian Basin created a small oasis of refuge for oilfield workers performing their daily tasks in a most inhospitable environment. (Photo courtesy The Petroleum Museum, Midland, Texas, Will Thompson Collection)*

company housing that catered to a more settled workforce. By then it was a common practice for the major companies to provide health insurance, stock option/retirement programs, vacation with pay, and other amenities associated with modern labor relations programs.[11] In 1924 both Humble and Texaco even established free recreation camps for their employees in the Houston area. For those planning to stay in the oilfield these amenities provided a strong inducement to go to work for one of the majors. But to many of the contact workers who had become addicted to the excitement of following the booms, that security was anathema. They viewed the company people who worked in roustabout gangs, who were pumpers or gaugers, or who operated the pump stations along the pipelines as somehow lesser beings who shied away from the real action, excitement, and opportunity of what the oilfield was all about. They refused to be tied down and lose their freedom of movement and independence of action.[12]

The best known of those boomers were the drilling crews. During the first twenty years of oil well drilling in Texas there was very little change in drilling equipment, and the crews became well versed in their use. The cable tools, which had been used since before the Civil War, had evolved into a reasonably efficient drilling system by 1900. They remained essentially the same until they went out of fashion in the 1930s. Rotary drilling rigs that were first used at Corsicana in the mid-1890s experienced minimal change during the same time span despite a general introduction of heavier-duty machinery and the 1909 introduction of the rotary cone bit. During those first twenty years or so of Texas activity the rotaries experienced limited use except on the Gulf Coast and other areas where there were unconsolidated formations at depths rarely below 2,000 feet. Consequently, there was not a great need for significant technological innovation in those rigs.[13]

The men who operated those drilling rigs were pulled from the surrounding rural labor pool and learned their trade through practical experience. It was generally accepted that it took about five years to develop a good cable tool driller. Their work required a significant amount of skill that was as much an art as it was an occupation. Their success in drilling depended as much on how the bit sounded when it struck at the bottom of the well, how the tension felt on the drilling line, and a host of other nuances that they learned to observe over a period of years. Those first cable tool men came to Texas from the old Pennsylvania and West Virginia fields and gradually trained new drillers from the cadre of raw country farm boys who went to the oil patch to make good wages. With the demand so high for experienced cable tool men from out of state, one Texas driller said in derision that "every tool dresser in West Virginia threw his sledgehammer in the Mississippi River when he crossed it and declared himself a driller when he got to Texas."[14]

The rotary men, by contrast, were pretty much a homegrown lot. Nobody knew much about the rotary drilling process when it was first introduced in Corsicana, so there were not a host of built-in prejudices toward change to overcome in doing the work. During that first twenty years or so the technology on those rigs remained very basic, and the drillers, like the old cable tool drillers, utilized a significant amount of native intelligence tempered by experience to operate their rigs. Consequently, they became notorious for their independent attitude as they were generally left to their own devices on how to accomplish the work. For example, when hoisting pipe out of the hole great strain was put on the derrick, which sometimes was "pulled in" or collapsed. Drillers learned to judge the amount of strain on the rig from the creaks and groans of the wooden derricks. In some cases they even devised a primitive form of a weight indicator by hanging a large section of pipe in the derrick so it lacked one foot reaching the drilling floor. As the strain increased on the derrick it would "squat" and the pipe would move downward. It was generally considered that eight inches of squat was all a derrick could stand before it collapsed, so all the driller had to do was not to exceed that limit.[15]

The Hamill brothers used a three-man crew to drill the discovery well at Spindletop, and it is unclear how many men constituted a rotary drilling crew in those first days. However, within a short period of time a five-man crew became standard and remained that way thereafter. That crew consisted of a driller, a boiler fireman, a derrick man, and two floor hands. With the exception of the driller they were collectively known as roughnecks, and in emergencies they could be used interchangeably, although the boiler fireman and the derrick man required a significant amount of experience. It was considered unsafe not to keep a well-trained worker on the job. The cable tool rig required only two workers, the driller and the tool dresser, or "toolie."[16]

The workers who operated the drilling rigs in those early days had an educational level that rarely exceeded the third or fourth grade. It was not unusual for some drillers to be illiterate and depend upon some other worker to fill out their drilling logs. Gerald Lynch relates that as late as 1925 when he went to work on the rigs he was the only member of the crew with a high school education, and the rest of the workers could barely read and write. With a workforce of that caliber most of those in the drilling industry were extremely suspicious of scientific or technological innovation that might affect the way they did their jobs. There was even a saying that developed in the patch that you should never tell a new hand anything because he might get your job. As a consequence, as long as the well depths did not exceed the 3,000 to 4,000 depth range of the old equipment they were accustomed to everything performed very well and there was little need for innovation from the point of view of the workers.[17]

Steel drilling derricks had been available prior to 1910, but once again they were not adapted for general use until the late 1920s and did not become the norm until after 1930, due in part to a reluctance on the part of the workers to change. The rotary men felt that they were dangerous because it was difficult to judge the stress on steel derricks, and the cable tool men felt that the wooden derricks had more spring to them, which contributed to a more efficient drilling operation. Those were only a few of the objections offered, but there were many others. One drilling contractor even maintained that it was easier to clean up after a fire on a wooden drilling rig; this was why he preferred them.[18] By the time of the East Texas boom in 1930 steel derricks were the norm. The obvious dangers of wind toppling derricks, fire, and a general high maintenance cost ultimately doomed the wooden derricks. Additionally, the 6,000- to 10,000-foot wells becoming common in the Permian Basin of West Texas mandated larger and stronger equipment, which only steel could provide.[19]

As drilling across the state proceeded in the late 1920s and on into the 1930s it became more and more obvious that the cable tool rigs could not drill to the depths required with any degree of efficiency. Inventions like the tri-cone rotary bit in 1933 and the weight indicator at about the same time, as well as a host of more powerful types of equipment, made the rotary rig the principal drilling machine by around 1940. As that equipment became more powerful and

*Coming out of the hole with a treble stand during the process of making a trip (removing the drill pipe on a rotary rig three 20-foot joints at a time in order to change the drill bit), ca. 1946. Note that in that era of changing safety procedures most of the roughnecks wear hard hats—they are dressed in overalls or coveralls—and they are wearing steel-toed work boots (Red Wing brand if they can afford them). (Photo courtesy of Southwest Collection/ Special Collections Library/ Texas Tech University, C. C. Rister Collection)*

complicated the need for a better-educated worker developed until by the time of World War II (ca. 1941) the day of the self-taught undereducated driller was rapidly drawing to a close.[20] By that time safety concerns of the major oil companies had pretty well mandated steel safety hats, steel-toed work boots, safety guards on moving machinery; a three-shift eight-hour tour had replaced the old two-shift twelve-hour tour that drilling crews had used since 1901.[21]

When all the steel rationing and restrictions on manufacturing lifted following the end of World War II there was a dramatic surge in producing new oil well drilling equipment. It was a time when the old steam-powered rigs were discontinued and a variety of new drilling machinery, much of it on the drawing board prior to the war, came on the market. It was an era that saw the wholesale adoption of internal combustion engines as a power supply and the development of a host of new improvements that allowed drilling wells to the 20,000-foot level or greater. During the immediate postwar period engineers, geologists, and a variety of other university-trained professionals began to supervise well drilling, and technological expertise became a necessity for the workers on the drilling rigs. The old boiler fireman became the motor man and the derrick man assumed the additional role of checking the specialized mud used for deeper

*Derrick man racking trebles coming out of the hole while working from the treble (monkey) board (derrick man placing the 60-foot stands of drill pipe between the fingers of the pipe rack during process of removing drill pipe from the hole while standing on a platform generally termed the monkey board), ca. 1946. If he is handling 60-foot lengths of pipe the platform is sometimes called the treble board, and if he is handling 80-foot lengths it is called the forble board. Note that the derrick man is not wearing a hard hat but is using a safety belt, although many of them disconnected the belt in order to have more freedom of movement. (Photo courtesy Southwest Collection/Special Collections Library/Texas Tech University, C. C. Rister Collection)*

drilling. The obsessive independence of the drillers disappeared in that new age. The driller, whose practical experience remained a valuable asset, remained the most important member of the crew, although he depended upon geologists and engineers to guide him in his work and he took his orders from the tool pusher who was in overall control of the rig. The two floor hands still worked the lead tongs and the backup tongs, but their job was made much easier by a variety of pneumatic equipment that did all the heavier work. By the 1960s most of the old-time hands agreed that the average crew of the 1920s and 1930s would not be able to operate the drilling rigs of the era.[22]

The changing face of the drilling industry directly affected other aspects of the oilfield contracting business. Most notably was the rig-building trade. The tough, hard-working and hard-living rig builders of those early wooden derricks were forced to change with the times. Between 1900 and the mid-1920s rig builders "laid out" their rigs with little more than the ubiquitous rig ax that was the distinctive tool of their trade. By the 1930s the rig ax had been traded in for the "spud wrench," an alignment and tightening tool used to erect the bolted steel derricks that appeared. Rig building was the only oilfield trade in Texas that embraced the labor union movement at an early date. Because of their relatively small numbers and the specialized labor they performed they were essential to keeping the drilling process going. After all, a well cannot be drilled if there is no rig. As a result the Panhandle strike of 1927 and the West Texas strike of 1936 by the rig builders temporarily shut down drilling operations until a settlement was reached. Those efforts rank as the only successful incursion of organized labor into the Texas oilfields since the failed attempt in 1917.[23]

In the post–World War II era when the drilling industry began manufacturing rigs again major changes soon developed in the rig-building trade. The bolted standard steel derrick quickly began to be replaced by unitized rigs that did not have to be assembled but could be moved by trucks. Although some rig building continued to be done on into the 1950s the bulk of the rig-building activity consisted of skidding the old standard derricks from one location to another. After a time all the older derricks were replaced by the newer "jackknife" type of unitized rigs with cantilevered masts; by 1960 the rig-building trade was a thing of the past. Thus, for a period of almost sixty years one of the most storied professions in the Texas oilfields maintained a presence that ultimately succumbed beneath the onslaught of technological change.[24]

Another group of oilfield workers with a particularly tough reputation were the tank builders. John S. Cullinan brought the first tank builders to Texas from the Pennsylvania oilfields in 1897 when he had at least two 37,000-barrel riveted steel tanks built to hold 100,000 barrels of contracted oil at Corsicana.[25] Cullinan was also responsible for building a large number of similar vessels to contain the Spindletop production until it could be processed by the new refineries at Port Arthur or loaded on railroad cars and oil tankers for transporting crude oil to other markets. That construction project produced the first tank farms in the state.[26]

The massive amount of oil produced on the Gulf Coast in those first days created an oil storage situation at the well sites that soon became a common practice throughout the region. It began with the discovery well in 1901 where dikes were

*Rig builders putting the finishing touches to a wooden derrick under construction on the McElroy Ranch in Crane County, ca. 1927. (Photo courtesy The Petroleum Museum, Midland, Texas, B. E. Hartwell Collection)*

thrown up to contain the free-flowing oil.[27] By 1903 the construction of those open earthen storage tanks was one of the few areas of oilfield work available to African American labor. It was primarily an earthmoving job done by scooping out holes in the ground, building dikes around the depression, and tamping the soil on the bottom and sides to prevent undue seepage. Those pits ranged in capacity from as little as 50,000 barrels to as much as 100,000 barrels.[28]

Within a few years the practice was standardized into a system described as:

> The oil runs into an earthen pit, from four to six feet in depth, with an area and capacity depending on the volume of production of the well. From this pit the oil is run into measuring tanks, and from the latter into the gathering or pipeline.[29]

The use of those earthen storage tanks was sometimes modified by making the inside edges of cypress wood with an earthen embankment reinforcing it from outside. Regardless of the construction technique, the Gulf Coast oilfields became dotted with what amounted to thousands of small lakes of oil as the size of the open tanks grew to 350,000- and even to as much as 500,000-barrel capacities.[30] By mid-1917 open pit oil storage on the Gulf Coast was credited with creating a 2 percent to 8 percent seepage, which amounted to a loss of 8,000,000 barrels of oil per year. It was recommended that oil producers replace the earthen storage with steel tanks capable of retaining all petroleum vapors. Because steel had become so expensive due to the demands of World War I it was suggested that concrete tankage buried below ground level with either wooden or concrete decks would be an acceptable alternative.[31] Less than a year afterward the Bureau of Mines issued a bulletin that recommended essentially the same solution to the problem.[32]

The use of those massive open storage pits continued to some extent until 1919 when representatives of the Texas Railroad Commission made an official visit to the Desdemona area to inspect oil wastage there and issued a general ban on the practice of storing light oils in open earthen tanks.[33] The use of giant field storage facilities continued with the introduction of concrete tanks with decks as recommended by the various studies. In 1917 the Empire Pipeline Company completed a 350,000-barrel concrete tank at Gainesville, Texas, in six weeks' construction and three weeks' curing time, and started building another soon afterward.[34] Soon firms like the Portland Hydraulic Tank Company of Shreveport, Louisiana, and the L. J. Mensch Company of Chicago with their reinforced concrete storage tanks for oils became active in building the new storage facilities.[35] Steel tanks of various large capacities were also advertised by a number of companies, like Rosenburg-Rowan of New Orleans, and steel construction became much more common. In 1918 a 55,000-barrel steel storage tank was built

*Typical earthen pit oil storage that dominated oil storage in Texas from its beginnings in 1901 until as late as 1920. (Photo courtesy Southwest Collection/Special Collections Library/Texas Tech University, C. C. Rister Collection)*

at Ranger, and by 1919 more than 4,750,000 barrels of pipeline oil was utilizing steel storage in the Ranger area.[36] However, as late as the mid-1920s two one-million-barrel concrete tanks were built in West Texas and one in the Panhandle, although they were never used. By that time large riveted steel storage tanks were the norm for the oil patch, and the large earthen and concrete storage facilities were only a memory of times gone by.[37]

The large, riveted steel tanks in the 55,000- to 180,000-barrel range ceased being built for general oilfield storage in the 1930s. During that time they were replaced by welded tanks of the same general capacity but concentrated into tank farms that supplied the pipelines and refineries with feed stocks. As early as the mid-1920s all the large tanks were equipped with floating decks that rose and fell with the amount of oil in the tanks. This greatly reduced the amount of space where gas could accumulate and consequently decreased the danger from explosion and tank collapse.[38]

Although the giant dirt, concrete, and steel storage vessels attracted much attention, they were not the focus of the tank-building trade as practiced by traditional oilfield hands. Both the dirt and the concrete tanks fell more into the earthmoving and general construction categories of labor. The large steel tanks, riveted early on and welded at a later date, fall into the tank-building trade albeit somewhat specialized in nature, but by the 1950s they had largely ceased being built due to the large capacities of the existing tank farms. It is generally conceded that those workers who erected the smaller wooden and bolted steel tanks used for temporary or field storage throughout the oilfields are best known as tank builders or "tankies."

With the start of the Texas oilfield work in 1901 the bulk of those field tanks were constructed of cypress wood in capacities ranging from 100 to 1,200 barrels. Wooden tanks continued to be generally used in the Gulf Coast region and throughout the Ranger, Burkburnett, and Desdemona booms on into the 1920s. Those vessels were fastened together with flat steel bands around cypress "staves" to create a tight seal, although the wood construction suffered considerably from seepage. By the early 1930s cypress began to be replaced with redwood held together with round threaded rods that were capable of being tightened at later dates in order to reduce the seepage problem. During the 1930s the wooden tanks were largely replaced by bolted steel tanks in 65- to 10,000-barrel capacities. By the 1940s steel totally replaced wood except in areas where corrosion was a significant problem.[39]

The builders of bolted steel tanks and the rivet crews of the large steel vessels were the workers who gave the trade its rough reputation. Bolted tanks made their first appearance in the oilfields around 1910 when both the Columbian Tank Steel Company of Kansas City and the Parkersburg Rig and Reel Company of Parkersburg, West Virginia, introduced them.[40] By the end of World War I several other firms, including the Maloney Tank Manufacturing Company; Black, Sivalls, and Bryson Tank and Manufacturing Company; and the National Tank Company, all of Oklahoma, had entered the fray and thousands of tank builders were at work in the Texas oilfields.[41]

The work of building bolted steel tanks was one of the most arduous in the oil patch. The tanks were constructed of five feet by eight feet steel side sheets called staves and the bottoms and decks were made of the same gauge steel. The sheets had to be carried by hand to the proper location and hoisted by block and line if the tank exceeded eight feet in height. Then the sheets were bolted together with half-inch bolts spaced on two-inch centers. A typical high five (a 500-barrel tank sixteen feet in height) would contain approximately 3,500 bolts that had to be tightened using hand-operated ratchet wrenches. "Peewee" Johnson stated that when he entered the profession in 1944 they were still using ratchets that caused his hands to swell to the point that his wife would have to cut his eggs for him in the morning because he could not hold a knife and fork. Those field vessels were normally constructed in batteries of two to ten tanks each dependent upon how much oil was being produced. Even with the introduction of pneumatic impact wrenches for bolt tightening in the late 1940s the work continued to be backbreaking labor.[42]

*Tank builders constructing a battery of 250-barrel tanks, ca. 1950. One tank has the staves (side sheets) standing, but not nutted out, and the tankies have finished laying the bottom on the second and are preparing to stave it up. (Photo courtesy The Petroleum Museum, Midland, Texas, Robert A. Estes Collection)*

By the end of World War II the large tank companies generally operated one four-man wood crew and several four-man steel crews from each of their branch locations. By the mid-1950s welded steel tanks, which were constructed in fabricating plants, began to replace bolted tanks, and by the 1970s practically all the new tanks used in the oilfield were either welded vessels or the noncorrosive fiberglass tanks that had replaced the wood tanks. By 1980 the tank builder, like the rig builder, was another oilfield trade that fell as a casualty to advancing technology.[43]

During the first twenty years or so of the petroleum industry in Texas pipelining was another one of those oilfield trades that employed huge numbers of unskilled workers for the general dirt moving aspects of the work. As a result it was one of the very few occupations in the oilfield that was available to the African American and Hispanic labor force, although by the late 1920s they were also largely excluded from the business. Additionally, the 1920s was an era that experienced large-scale technological change in the way pipelines were laid.

One of the last major traditional screw-type pipelines laid in Texas took place in 1927 when the Gulf Oil Company built 546 miles of combination eight- and ten-inch line from Crane in West Texas to Lufkin in the southeastern part of the state. That job employed mechanical ditching, laying, and filling machines instead of using hand labor, and they averaged four and one-half miles per day instead of the three-quarters of a mile average of twenty years earlier when all the work was done by hand. Despite those innovations it still required six gangs of 150 to 200 men per gang to lay the line, and they were still being housed in portable tent camps just as those 1905 pipeliners were housed in the first days of Texas oil.[44]

By the time the Crane to Lufkin line was completed welding had already entered the pipelining trade, and further changes were well under way. During the 1923–24 period acetylene welding was becoming used on a regular basis in the natural gas industry, as exemplified by the line the Empire Natural Gas Company laid that held three hundred pounds of pressure with no leakage.[45] In 1925 the Magnolia Gas Company completed the first major acetylene welded pipeline using fourteen-, sixteen-, and eighteen-inch pipe on their 217 mile line from northeast Texas to Beaumont, making the process well accepted for pipeline use for several years following that job. By the early 1930s acetylene welding began to be replaced by the electric welding process, which, along with the introduction of seamless pipe, heralded significant changes in the industry that were destined to remain essentially the same for the rest of the century.[46]

Those changes that the pipelining trade experienced during the early to mid-1930s reduced that industry's labor needs by approximately 75 percent. The ditching was done by machine, which required minimal labor beyond a trained operator. Connecting the pipe was done by welders, which was a skilled occupation that mandated significant training and eliminated large numbers of untrained workers. Laying the line in the ditch was done by machine, which reduced that labor force by 50 percent. Finally, the backfill work was also done by machine, which also required a trained operator and a small force of unskilled labor. Thus, by the period of World War II pipelining had experienced significant technological change and the hordes of workers that once followed that trade were greatly reduced to a core of trained professionals with specific specialities.[47] With that change in focus by the 1960s a significant number of pipeliners opted to join the organized labor movement.[48]

Prior to 1920 the oilfield was largely dependent upon animal-drawn vehicles for its immediate transportation needs. Railroads transported the freight to the depots nearest the site of the latest oil strike, and from there the teamsters did the freighting. Generally speaking the distances covered by those operations were less than twenty miles. About the time of World War I it became recognized that gasoline-powered trucks were the key to the future of the oilfield freighting business in Texas due, in part, to the greater haul distances required by many of the new discoveries. Despite the obvious superiority of motor truck transportation the road infrastructure of the state was sorely lacking.

During the first decade of the twentieth century Texas roads were little more than cleared and graded pathways, and the bulk of all freighting and travel was handled by the railroads. Any cross-country freighting in that era was of such light weight that the existing roadways seemed adequate. But as early as 1903 "good roads" associations began to be formed across the state when it became apparent that better roads were going to be needed as automobiles became more plentiful and cross-country travel became more popular. At that time all road building was under the purview of individual county governments, and it was difficult to get a statewide consensus toward building better roads. In 1917 the Texas State Highway Department was established to take advantage of funds made available under the Federal Aid Road Act of 1916. Accordingly the state legislature authorized a system of highways that would be built by the various counties coordinated by and under the supervision of the Texas Highway Department. It was not until 1925, however, that the highway department took full control of building the highways in the state. This marks the beginning of a good system of roads in the state that benefited the trucking industry on a large scale.[49]

The petroleum industry recognized the need for motorized transport in the oilfield and was an avid supporter of the movement toward better roads in the state. During 1918, in the midst of World War I, the trucking industry made great strides in handling oilfield freight, such as the trailer system developed on the Gulf Coast that was capable of transporting all the materials for building a complete wooden drilling rig in one trip as opposed to the several trips required when using wagons and teams.[50] Then, as heavy oil development took place in the Wichita Falls and Ranger areas, the value of using trucks became more and more evident. Trucks became the primary mode of transport because of the greater cost of teaming due to the longer hauls done in those areas. It was estimated that 95 percent of the freight hauled from the rail yards at Wichita Falls to the scene of the boom at Burkburnett was accomplished by motor truck.[51] Because most of the hauls around the Ranger area were twenty-five to sixty miles in distance, trucks became the major freight hauler in that boom. By late 1918 there were 350 trucks operating in the Ranger area and 1,400 teams and wagons. However, the overwhelming amount of freight was handled by trucks, while the teams were largely relegated to excavation work for tanks, pipelines, and other dirt work. But even then there was beginning to be a growing preference for using tractors instead of horses in the dirt contracting business.[52]

It was generally accepted that as soon as the war ended the combination of more powerful trucks developed for the war effort combined with renewed domestic manufacturing would bring a generation of much better trucks to the oilfield. This also called for better roads in the areas of oil production because it was a well-known fact that the graded gravel roads of the day could not stand the large-scale wear and tear created by the heavily loaded oilfield trucks that crowded the roadways. Consequently, plans were developed for concrete highways in the developing oilfields of the central West Texas region in Eastland County.[53]

Following the war, as oil development moved into the Panhandle and the Permian Basin regions, motor trucks became the dominant mode of freight movement in the oilfields. The longer distances covered and the need for fast transport quickly relegated the wagons and teams to a minor role. At the height of the Borger boom in mid-1926 practically all the freight hauling was done by motor truck.[54] Meanwhile, out in West Texas much of the early freighting, beginning in 1923 at Big Lake and continuing on into the Wink boom in the late 1920s, was done by horse-drawn vehicles. But as soon as the contractors learned how to deal with the dry sandy environment the animal-drawn vehicles disappeared from the scene. By the time of the East Texas boom in 1930 the use of wagons and teams for work in the oil patch was largely gone despite the anecdotal comments of participants interviewed years afterward.

The colorful oilfield freighting business that utilized thousands of horses and mules is one of those professions that did not suddenly disappear, but gradually faded from the scene as automobiles and roads became more efficient. It held a special place in the memories of those who made the various booms; those memories in many ways connected those folks to the life they left when they ventured into oilfield work. It is interesting to note that the teaming work described by participants of the Gulf Coast boom of the 1900–1910 era is presented in a general matter-of-fact manner, as in "this is how we did it in those days." However, as time progressed the recollections fall more and more into the nostalgic vein of how the old-fashioned teams became the saviors of the newfangled motor-driven vehicles when the teams had to pull the trucks out of mud holes, sand beds, or some other spot where they could not go. There is little question that in the workers minds somehow the old rural background had triumphed in some small way over the machine age that had intruded on the workers' world. However, once again technology triumphed as the trucking industry managed to stay in step with the rest of the nation toward an ever more efficient operation of an industry.[55]

Contrary to the popular image, oil wells do not usually spew thousands of barrels of oil high into the air when they are completed. Most of them actually produce only a portion of the amount necessary for the well to make a profit when they are first completed. This is a result of the density or porosity of the formation in which the well is drilled, which will not allow much of the oil trapped between the pores in the rock to flow into the well bore. Therefore, it is necessary to stimulate the well to increase its flow and produce enough oil to show a profit. As early as 1865 oilmen discovered that by exploding a charge of nitroglycerin at the bottom of the well the explosion would crack the stone outward from the well bore and greatly enhance oil production. By the time of the discovery of oil in Texas at the beginning of the twentieth century the occupation of well shooter as an oilfield specialist was well established.[56]

Well shooting was not generally used in the soft formations of those first discoveries along the Gulf Coast, but by the mid- to late teens, as drilling activity moved farther and farther north, the harder limestone and sandstone formations found there required fracturing to get the best oil production. Consequently, a number of firms, usually headquartered in Ohio and Pennsylvania, who specialized in shooting oil wells began operating in the Texas oilfields. Due to the extremely dangerous nature of transporting nitroglycerin the product was manufactured and stored in close proximity to the oilfields where it was used, but always in isolated locations far away from towns and human habitations.[57]

The manufacturing process, called nitration, was a relatively simple process accomplished by slowly adding glycerin to a mixture of nitric and sulfuric acid to produce liquid nitroglycerin. The resulting product was poured into square metal cans holding ten quarts and weighing slightly more than thirty-three pounds. The cans were stored in wooden or corrugated metal buildings called magazines located a mile or more from the nitration plant. Normally each magazine contained approximately one thousand quarts of the explosive material. It was necessary to closely monitor the temperature at the nitration plant and the magazines because nitroglycerin becomes very unstable when frozen or is subjected to temperatures in excess of 118 to 120 degrees. Nitroglycerin is also an explosive capable of withstanding jolts of up to twenty-eight pounds per square inch, which makes transport reasonably safe if proper care is taken.[58]

By the 1920s shooting wells in the north and north central part of the state around Wichita Falls, Ranger, and Breckenridge was a common practice. As activity moved into the Panhandle and West Texas the amount of explosives necessary to shoot a well grew from the one hundred–quart level to a thousand or more quarts per well depending upon the well depths and the nature of the formations being shot. During those days the shooters all used automobiles to transport their product from the magazines to the well sites. For obvious reasons they preferred the larger vehicles like Buicks, Cadillacs, and Pierce Arrows that had better suspension and gave smoother rides. A shooter's transport car was equipped with a flat space in the rear where the trunk was normally located. That space was equipped with a series of rubber- and felt-lined vertical bins exactly the size of the ten quart nitroglycerin cans so that when loaded there was a minimum of jostling. Most of those cars had between fifteen and twenty such bins which allowed them to transport between one hundred fifty and two hundred quarts of the explosives. By the late 1920s, as the deeper West Texas wells required as much as 2,000 quarts per shot, the shooters began to use trucks for transport to those larger jobs.

When a shooter arrived on the job he first checked the well to ascertain the exact depth and to make sure there were no downhole obstructions. Then he filled each torpedo or shell (the elongated tin containers with soldered joints that held the nitroglycerin for the shot) with water to check for leaks. Once the shell was found to be leak-proof it was hung by a hook attached to a steel line above the well opening to be filled with explosives. Before the nitroglycerin was removed from the car the route to the well opening was carefully cleared of all possible obstructions and other possibilities of slipping. Then the ten quart containers were taken up on the rig floor two at a time, so the shooter could maintain good balance, and carefully emptied into the shell. When the shell was full it was lowered to the bottom of the well and the process was repeated until the requisite amount of the explosive rested at the bottom of the hole. Afterward the entire batch was exploded and the formation was fractured enough for the oil to flow.[59]

Until the mid-1920s the primary way the shot was ignited was through the use of "jack squibs" which consisted of a stick or two of dynamite with a two minute fuse that was lit and dropped down the hole. The general unreliability of that method precipitated the development of a variety of time bombs such as the so-called Bolshevik bomb that was simply an alarm clock attached to an explosive device that was lowered atop the nitroglycerin. It exploded when the alarm went off. By the early 1930s that system was replaced with much more reliable electrically ignited bombs such as the Zero Hour Bomb manufactured in Tulsa, Oklahoma.[60]

By around 1940 the liquid nitroglycerin began to be replaced by a gelatin nitroglycerin with the general consistency of Jell-O that was much safer to transport and handle than the traditional material. The gel was manufactured by large concerns like Dupont and was shipped to the shooters like any other product. This negated the need for the nitration facilities scattered across the oilfield.

*Preparing a well to be shot in the 1920s. Note the shooter pouring nitro into the shell, his helper adjusting the measuring device for determining the depth the shell is lowered, and the compartment on the back of the truck specially designed for transporting nitro. (Photo courtesy Panhandle-Plains Historical Museum, Canyon, Texas)*

Despite the increased safety of the new product it was still dangerous and the magazines remained in their isolated locations at a storage facility for the explosives used by the shooters. C. L. McKinney declared that when he went into the shooting business for himself in 1945, gel was the only explosive he used for shooting. By that time the practice of shooting oil wells with nitroglycerin was a rapidly declining occupation due to the general adoption of sand fracturing and acidizing developed by Eddie Chiles and others five or six years earlier. When McKinney went out of business in the mid-1950s the job of the well shooter, like so many others in the oil patch, was a trade no longer practiced.[61]

The romantic nature of such a dangerous occupation as oil well shooting has lent itself to a number of hair-raising stories. Most of them revolve around unanticipated explosions that leave only bits and pieces of unfortunate shooters and innocent bystanders to explain the exact nature of the mishap. Both C. L. McKinney and Toby Mendenhall aver that those incidents were exceptionally rare, and were almost always the result of carelessness on the part of the shooter. McKinney in particular was convinced that all the transport explosions were the result of undetected container leakage aggravated by friction and never from the unexpected jars and bounces of the transport vehicles. Both men also agreed that most of the premature well explosions were the result of high downhole well temperatures that were little understood in the days of the well shooter.[62]

However, the most celebrated stories associated with shooters deal with how some of them captured loaded shells as they were forced up out of the well by gas pressure. There are a number of accounts of those events, and even one of an unsuccessful try in which the shooter survived the resulting disaster. Mody Boatright investigated several of those and declared most of them eminently worthy of being taken with a grain of salt. Nevertheless, at least two of the accounts, that of Walter Cline's witnessing Tom Mendenhall accomplish the feat sometime before 1918 and Ed P. Matteson observing the same with another shooter a couple of years later, appear to be the only credible examples of such a remarkable activity.[63]

An offshoot of the well shooting occupation became that of the oil well firefighter. Oil well fires posed one of the distinct oilfield dangers beginning with the first discovery at Spindletop and lasting until the present. There is credible evidence that the first oil well fire extinguished by explosives was accomplished about 1908 in southeast Texas by a Texas Company superintendent named C. M. Chester.[64] However, there are competing claims, such as that of a California shooter named Ford Alexander and the account of a similar incident near Veracruz, Mexico. Myron Kinley, the father of modern oil well firefighting techniques, even claimed to have put out a well fire in California as early as 1913. Regardless of who accomplished the feat first, by around 1920 shooting out oil well fires with explosives had become relatively common as a means of last resort in well firefighting.[65]

Oil well fires are always caused by the ignition of natural gas at or near the wellhead. For that to happen there has to be the proper mixture of gas and oxygen within well-defined percentages. If there is too much gas it will not burn and if there is too much oxygen it will not burn. The purpose of the explosion is to deprive the flame of oxygen, changing the mixture and extinguishing the fire. During the 1920s and earlier the same result was accomplished by setting up a number of steam boilers and smothering the fire with a series of steam hoses concentrated upon the fire. That process, if successful, took from a few days to a week or more and required the services of thirty or more workers. If the fire was shot with explosives the result was the same within a much shorter time frame and at a considerably lower cost to the well owner.[66]

Given the dangerous nature of oil well firefighting, very few men became involved in the process on a regular basis. If they survived any length of time they usually became celebrities of a sort. Ward "Tex" Thornton was one of those who achieved such fame. Arriving at Electra in 1920, Thornton soon achieved somewhat of a local reputation as a firefighter although his principal job was that of a shooter. In the mid-1920s he moved to the developing boom in the Panhandle and headquartered at Amarillo. He built up a good business as a well shooter and put out several fires over the course of a couple of years. Then in 1927 he extinguished three oil well fires in the course of two weeks and was hailed in the press as the reigning king of oil well firefighters. The following year he extinguished a major fire near Corpus Christi, where he used an asbestos suit that he claimed he invented. From then until his death in 1949 the headline-seeking Thornton maintained a public persona as the daredevil of the oilfields.[67]

Although Tex Thornton received a great deal of press in his lifetime, a contemporary of his named Myron M. Kinley is credited with bringing modern oil well firefighting to fruition. Born in Pasadena, California, in 1896, he began his firefighting career while still a teenager in his home state. He later moved to Oklahoma where he gained a considerable reputation as a firefighter and well control specialist. One of the most overlooked aspects of oil well firefighting is capping the blowing well and bringing it under control after the fire is out. That is probably the most hazardous part of the business, and it is actually the ultimate purpose of the entire operation, although the fire is the most spectacular part. Kinley burst on the public scene in 1931 when he extinguished a fire in Romania that had been burning for 890 days and was considered impossible to control. Shortly afterward he formed the M. M. Kinley Company, which specialized in shutting down oil well blowouts and putting out fires.[68]

Perhaps the most famous of those oil well firefighters is Paul Neal "Red" Adair, who went to work for Kinley in 1946 and formed his own firefighting firm, the Red Adair Company, Inc., in 1959 in Houston. He, like Thornton and Kinley before him, captured headlines for years with his spectacular exploits although he was a stickler for safety and developed numerous technological innovations that built on the work of those who had gone before him. Unlike the job of the shooter, the work of the oil well firefighter has remained an important part of the industry.[69]

All those oilfield occupations that began in the earliest days of the industry in Texas reflected the need of some aspect of the industry. At first the response was rather crude due to the technology of the day. Given the relative simplicity of the machinery used to find, produce, and transport petroleum, the local rural population was able to fill the labor needs almost instantly. As time passed and oil industry technology grew more sophisticated the labor force changed to fit the changing nature of the work. Those young men accepted the hardships of living and working in boomtown environments as an adventure that paid them well. Over time that aspect of their life also became easier as transportation developed and changed the aspect of the boom into a more settled situation. The result was that over a forty-year period the work changed into a complicated technology and the lifestyle changed from a frontierlike existence into more of a settled lifestyle.

The way labor was accomplished and the lifestyle of oilfield workers in Texas changed dramatically over time. When oil was first discovered the state had an overwhelmingly agrarian population. Most work was accomplished by hand labor and transportation was primarily animal powered. Over the following four or five decades the plentiful energy source provided by the oil industry created a technological revolution that changed both the labor and the lifestyle of oilfield workers. In the area of work mechanization, directly related to the development of the internal combustion engine, the numbers of pipeliners was reduced by well over 50 percent while efficiency was increased by several hundred percent. Mechanization totally changed the drilling industry by providing the power to drill deeper and faster than ever before, and in general affected every aspect of oilfield work. In the area of lifestyle the changes in transportation had a significant impact. In the days of animal transport workers had to live adjacent to their workplace, with the result that boomtowns developed and disappeared with alarming frequency as oil was discovered in various locations. As internal combustion–powered vehicles became more efficient the need to live close to the work site disappeared. By the time mobile homes became available in the 1940s the boomtowns of the 1920s became totally unnecessary.

As the Texas petroleum industry expanded over time its work force consistently mirrored the attitudes of the general populace of the state. It consisted of an overwhelmingly male Anglo group whose members reflected societal attitudes of the day in relegating the African American and Hispanic populations to a very minor role. In its earliest days on the Gulf Coast the industry only employed African Americans in earthmoving capacities such as digging slush pits, tank excavation, pipeline trenching, and other similar low-paying occupations. By the 1920s, with the national upswing of the Ku Klux Klan movement, the African American population was totally excluded from any oilfield labor on the Gulf Coast, although during the East Texas boom in the 1930s African American labor was used in a limited capacity in excavation work. In general the African American population was totally absent from mainstream oilfield work throughout Texas. They were relegated to dirt work in the early days and later worked in the Gulf Coast refineries in low-paying capacities such as custodial work.[70]

Similarly, the Hispanic population was largely absent from oilfield work except on a limited regional basis in the South Texas region centered on Laredo, where they performed similar labor. That minority group, much like the African American population, also found employment in lower-level jobs in the Gulf

*A Hispanic pipeline crew preparing a ditch for a pipeline in South Texas during the 1920s. (Photo courtesy The Petroleum Museum, Midland, Texas, J. D. Bonner Collection)*

Coast refineries at an early date. By the advent of World War II they represented a sizeable group in that area.[71]

In addition to the exclusion of minority employment oilfield workers held a general anti–labor union bias. Although the rig builders seemed to have embraced unionization, other oilfield occupations, particularly the drilling industry, never gave it more than a passing glance. The transient nature of almost all the contract labor in the oil patch mitigated against unionization of any type. Their independent mind-set combined with the unsettled nature of the work that depended upon a boom and bust cycle worked against unionization. In their mind having a union only saddled them with another entity of control. From their point of view if they were dissatisfied with the job they could just whip the driller and move on to another job that would probably pay more money anyway. The oil company workers, who most fitted the description of those ripe for unionization, also rejected the temptation. Most of them were recruited from the ranks of the boomers and they came to their new job with a built in anti-union bias. That, combined with the paternalism of the companies who provided housing, health benefits, and better-than-average wages, dissuaded them from joining the union. In general in the years following World War II oilfield workers remained union free with the exception of the refinery and petrochemical workers on the Gulf Coast.

Those then are the oilfield workers of Texas. They began their life in the oilfield as boll weevils and if they stayed they became hands. Some of their occupations, like rig building, tank building, and well shooting, have disappeared beneath the onslaught of technological change. Their labor and living conditions have changed significantly over time due to that same technological change, but through it all they have remained a remarkably independent lot who eschew organized labor as a restriction of their independence. They have a reputation for hanging out in beer joints and other disreputable establishments where they have been known to become somewhat rowdy. And as a result they have gained the somewhat tarnished moniker of oilfield trash. But at the end of the day they are those hands in greasy overalls who made the boomtowns boom and the oil patch work and they have provided an enduring legacy for the Lone Star State.

CHAPTER 10

# A Language of Their Own

Every industry has its own specialized vocabulary that gives insight into the nature of that industry. The petroleum industry is no exception to that rule. The language of the oil patch is colorful to outsiders and useful to insiders. Some of it harks back to the rural roots of those early-day participants, but most of it derives from the specialized equipment and the work associated with that equipment. Because there is such a diverse group of specialized trades within the industry, such as well drilling, rig building, or tank building, there is terminology directly related to those specific jobs as well as language common to the industry as a whole.[1]

## GENERAL TERMINOLOGY

Perhaps the two most commonly used terms in the oilfields of Texas are *boll weevil* and *oil patch.* Both date from the earliest days of the state's oil activity and both relate directly to the rural origins of the workers. *Boll weevil* is the term applied to an inexperienced worker, although beginning in the 1950s it morphed into the word *worm,* which has come to be used more and more by newer generations of oil workers. Boll weevil is used in a variety of ways, such as calling a mistake at work a boll weevil stunt or referring to being *weevil bit* when injured by some activity of the new guy. *Oil patch* is the general term used to denote the oilfield. The oil patch replaced the cotton patch as a workplace for those early-day workers; they many times simply refer to it as the *patch.* It is interesting to note that the term has come into general use across the nation when such unlikely persons as Wall Street analysts regularly refer to the financial situation of the oil patch.

API is another term used universally by oil workers. To the uninitiated it refers to the American Petroleum Institute under whose aegis much of the oilfield equipment has come to be standardized. However, to the experienced oilfield hand it means something is correct when he states, "That is just about API."

Another term that has gone out of vogue, but which all of the early-day hands used, was "I rode in here on the first load of timbers and I rode out on the last load of pipe." This refers to the fact that they were present during a particular boom. A term that means almost the same thing is the word *made,* which was also used from the earliest days of the industry and is still used today. The context for it would be something like, "I 'made' the such-and-such boom back in such-and-such a year."

*A Christmas tree, its name an example of the specialized terms adopted by oil patch hands for a specific piece of equipment. (Photo courtesy Southwest Collection/Special Collections Library/Texas Tech University, C. C. Rister Collection)*

A couple more universally used terms around the oil patch are *flanged up* and *drag up.* Flange up refers to the last step in pipe connection work when the pipe is connected to the flange. Hence, "flanged up" or "all flanged up" means that a job is complete. To "drag up" means that you quit a job and collected your money.

The word *the* is also used in an interesting manner among oilfield workers. When referring to oil companies, and particularly the majors, they almost always place "the" in front of the name. They might reference "the Gulf" or "the Humble," for example. That same usage is also used by cowboys when referring to ranches by their brands, although they always make the names plural, as in "the XITs" or "the 6666s." It is almost certain that this usage was transferred from ranch work to oilfield work when so many cowboys were lured into a new profession by higher wages.

*Location* is another word universally used by oilfield workers. In the strictest sense it means the spot upon which a drilling rig, a tank battery, or some other specific object is located while work is being done. However, in general usage it applies to someone being where an event happened, as in "I was on 'location' the night rig #8 caught fire."

The word *contract,* and variations thereof, is yet another interesting term. It applies to almost any oilfield work being done as having a "contract" on the job.

That does not necessarily mean there is any legally signed document concerning the work. It is also used to denote getting a job done in a fast and efficient manner by saying we are really "contracting" now. The origin of the term goes back to the earliest oilfield days when almost all work was done at a set price per day or for a particular short-term job under what was called a "contract" as opposed to an hourly wage basis. Under those conditions, if the job was finished in four hours or ten hours the worker received the same amount of money.

## DRILLING

In the mind of the general public oil well drilling has the highest profile of any oilfield occupation. Over time there have probably been more oilfield workers employed in that activity than any other. Consequently, a great deal of oilfield terminology relates to drilling, and some of it has even seeped into general oilfield usage. For example, "turning to the right" or "turning to the right and making hole" are generally used to denote getting the job done. The phrase refers to the activity of the drill pipe turning to the right when drilling is taking place.

Another phrase associated with drilling is "drilled past me and set pipe." For example, if someone stole a roughneck's girlfriend he might say, "He drilled past me and set pipe before I even knew he was on location." Then there is the term *twisted off,* which is the term used to describe a drilling accident when the torque on the drill pipe becomes so great that the pipe breaks. The hands use the term *twisted off* to describe someone making a serious error many times associated with going on a drunk and creating a disturbance.

One of the more interesting words associated with drilling is *tour.* "Tour," pronounced tower, is the shift worked by a drilling crew and it is not used by any other type of oilfield work. The term was in use as early as the late nineteenth century and there is no solid documentary proof of its origin. Tours consisted of two twelve-hour shifts (noon to midnight) until the 1930s, after which the workday was divided into three eight-hour segments. A phrase associated with "tour" in the days of the twelve-hour cable tool shift was "drilling against," as in, "Back then I was drilling daylights against Joe Blow." It meant that the speaker was drilling on the daylight "tour" (noon to midnight) and Joe Blow was drilling on the graveyard "tour" (midnight to noon).

## RIG BUILDING

Rig building, which finally faded out as a profession around 1960, has some specialized language that should be noted. For example, rig builders refer to the act of building a rig as *laying out.* In the early days of the industry back in West

Virginia and Pennsylvania, the wood for a derrick was "laid out" on the ground and the various parts were cut to size according to a pattern that generally conformed to the size at the bottom and the size at the top and the height. Those parts were "laid out" in a pattern on the ground and the derrick assembled according to that plan.

## TANK BUILDING

The boss of a tank building crew is called the *setter* because he *sets* or places the tank where it is supposed to go. Additionally, the term for building a tank is to *set* it, as in, "We set two hi-fives out on the ABC lease."

We had "the grade made and the bottom laid before those other hands got on the job" was a phrase tankies used to brag about how fast they worked. The term for getting a job done properly was "get it straight and turn it to the right," which refers to the act of putting the nuts on the bolts in a fast and correct manner.

## GLOSSARY

The following is a listing of words associated with the petroleum industry that are related to life and labor in general. All of these terms relate to specific lines of work or lifestyle in the oil patch and many of them are unique. I make no pretense that this is the most exhaustive collection of terminology of the patch, but it does give some indication of the extent and complexity of the language associated with an industry as large, diverse, and extending over such a long period of time as the petroleum industry.

**API**: The initials of the American Petroleum Institute, which sets the standards for most oilfield equipment, but the way it is used by the oilfield hands is to indicate something is correct. When they agree with something they might say, "That is about API."

**annulus**: The space between the drill pipe and the edge of the well bore during the rotary drilling process. Also referred to as the annular space.

**auger stem**: A section between the rope socket and the bit on a cable tool drilling string that adds weight to the string.

**back off**: Unscrewing the drill pipe by reversing the direction of the rotary table.

**backup tong**: Heavy-duty tong used to tighten or loosen the joints of a drill pipe. It is used in conjunction with the lead tong. The lead tong does the actual tightening or loosening and the backup holds the pipe still.

**bailer**: Device on a cable tool rig used to dip the slush created by the drilling operation from the bottom of the well.

**band wheel**: The ten-foot diameter wheel on a cable tool drilling rig that provides the power transfer to operate the rig.

**Barrett jack**: See circle jack.

**batter**: The slope or angle of a drilling rig derrick from the floor to the crown block.

**blowout**: An event that transpires when there is an uncontrolled expulsion of gas or liquid from a well.

**boiler man**: See motor man.

**boll weevil**: An inexperienced oilfield worker. The term has been used since the first days of the Texas oil industry to denote the rural roots of the new hands. Since the 1950s the term *worm* has been used interchangeably with boll weevil.

**Bolshevik bomb**: A device used in the late teens and early 1920s utilizing an alarm clock to explode nitroglycerin when shooting an oil well.

**bottomed**: See t.d.

**bowl and pitcher hotel**: Early day hotels without running water where many oilfield workers lived.

**boxcar house**: See shotgun house.

**breaking out**: Unscrewing drill pipe.

**bring in a well**: To successfully complete drilling an oil well.

**broke out**: Denoting time one began working in the oil patch. As in "I broke out during the Snyder boom in 1950."

**bull gang**: See roustabout.

**bull wheels**: A large set of wheels connected so as to create a spool upon which the drilling line on a cable tool rig is wound.

**cable tool drilling rig**: The type of drilling rig that creates an oil well through the percussion method of pounding up and down with a chisel-shaped bit suspended on a steel cable. It was the only type used in the mid-1800s and was used extensively until it went out of vogue in the oilfields around 1930.

**caliche**: A calcium carbonate stone used to surface oilfield lease roads in West Texas. A similar covering, usually termed chalk, is used in Central Texas while shell is used in the Gulf Coast region.

**camp**: A cluster of houses, usually in the midst of an oil field, housing oil company employees.

**cat**: The term used to denote an experienced pipeliner.

**cathead**: Spool-shaped attachment on the winch of the drawworks around which a rope is wound for hoisting and pulling.

**cat line**: The rope used on a cathead.

**cellar**: The pit below the drilling floor of a rig where various well control devices are installed.

**Christmas tree**: A series of valves and chokes installed atop a well to control oil and gas pressure. It received its name from the many projecting valves and pipe stems that give it an appearance similar to a Christmas tree.

**circle jack**: Steel-toothed semicircular track device attached to the floor of a cable tool drilling rig that utilizes a double action traveling jack for making up and loosening cable tool drilling tools.

**circulation**: The movement of drilling mud from the mud pit, through the drill pipe, and back to the mud pit during the rotary drilling operation.

**collar tapper**: The member of a screw-type pipeline crew who tapped on the end or collar of a section of pipe to establish the rhythm of the tightening crew. Also called a collar pounder.

**connection crew**: See roustabout.

**crater**: Large hole created by a blowout.

**crown block**: The device at the top of a drilling rig derrick that contains the sheaves or pulleys through which the drilling lines are threaded.

**cuttings**: Small chips of stone produced by the drilling operation.

**deck**: The roof of a tank.

**derrick man**: Worker on a rotary drilling rig whose primary responsibility is working with the drill pipe in the derrick.

**dog house**: A small structure associated with a drilling rig where the crew changes clothes and where small tools are stored.

**doodle bug**: In the early days of the petroleum industry one who used a divining rod or other instrument to find a place to drill for oil. At a later date the term applied to those working with seismograph equipment to locate likely places to drill.

**downhole tools**: Tools used inside the well bore during the drilling operation.

**downstream**: That portion of the petroleum industry downstream of the oilfield proper that includes refining and marketing.

**drawworks**: The machinery on a rotary drilling rig that operates the hoisting and drilling operations.

**drill collar**: Heavy drill pipe installed directly above the bit on a rotary drilling string that provides extra weight, which assists in keeping the hole vertical.

**drill pipe**: The pipe to which the bit is attached on a rotary drilling rig.

**driller**: The worker on a drilling rig actually responsible for drilling the well. On a cable tool rig he supervises a tool dresser, or "toolie," and on a rotary rig he supervises four roughnecks.

**drilling in**: Penetrating the production formation and setting casing during the drilling operation.

**drilling mud**: Material added to the drilling fluid in the rotary method of drilling in order to make the process more efficient.

**drilling rig**: The combination of the derrick, the drilling equipment, and the power supply on an oil well drilling machine.

**dry watch**: Acting as watchman over a stacked or stored rig.
**elevators**: The clamping device used on a rotary drilling rig that holds drill pipe or casing as it is raised from or lowered into the well bore.
**farm boss**: Person in charge of the operation of an oil lease. It derives from the early days of the oil industry when oil leases were called farms.
**finger board**: A set of racks in the derrick of a rotary drilling rig at the level where the derrick man works, designed to contain the drill pipe that is removed from the well bore.
**fish**: Foreign object lost downhole during the drilling operation.
**fishing**: The practice of removing foreign objects lost downhole during the drilling operation.
**fishing tools**: Those downhole tools used during the fishing operation to retrieve foreign objects lost in the well bore.
**floating deck**: The roof of a large tank that floats on the oil in the tank in order to reduce the amount of gas accumulation in the vessel.
**floor hands**: The two workers on the floor of a rotary rig responsible for connecting and disconnecting the drill pipe. One operates the lead tongs and the other operates the backup tongs.
**forbal board**: See monkey board.
**four ups**: The use of four horses abreast to pull freight wagons.
**gauger**: The pipeline employee who gauges or measures the amount of oil in a tank before it is transferred into the pipeline.
**geronimo line**: Emergency escape cable strung from the derrick man's workplace in the derrick some distance out from the rig as a means of escaping in case of some emergency.
**greasers**: Work clothes of oilfield workers.
**growler board**: Pipelining device placed beneath jack board in order to stabilize it during the pipe-laying process.
**headache post**: Beam placed upright on the floor of a cable tool rig to prevent the walking beam from falling on the workers in case it becomes dislodged.
**hijack**: Term for robbery popularized during the Ranger, Burkburnett, and Breckenridge booms in the late teens and early 1920s.
**hitch on**: The process of connecting the drilling line to the walking beam in preparation to begin drilling on a cable tool rig.
**hooks**: Pipeliner term for lay tongs or large pipe wrenches used to tighten screw pipe.
**independents**: Those smaller oil companies usually owned and controlled by an individual or a small group of individuals.
**jack board**: Device used by pipeliners to align screw pipe so it can be tightened
**jackknife rig**: A self-contained style of rotary drilling rig that came into popularity in the 1950s and has remained the norm to the present. Its value lies in

that it is capable of being moved from location to location by truck in a very efficient and time-saving manner. It replaced he standard rig used up until that time.

**jars**: A device that resembles gigantic chain links that attaches above the auger stem on a cable tool set of drilling tools. It provides a jerking motion to the process that increases the striking power of the bit.

**jug hustler**: Worker who puts the sensor units, called jugs, in place for a seismographic operation.

**kelly**: A heavy square or hexagonal device suspended from the swivel on a rotary rig that extends through the rotary table and is connected to the drill pipe. The rotation of the rotary table causes the kelly to turn, which rotates the drill pipe and creates the drilling process.

**killing a well**: Bringing a threatened well blowout under control by injecting a dense fluid into the well.

**land man**: See lease hound.

**layout**: Rig building term meaning to build a rig, as in, "I laid out three eighty-foot rigs for the ABC Company." The term derives from the early days of the oil industry when wooden rigs were laid out in a pattern on the ground and measured for size before construction.

**lazy bench**: A crudely made bench on the floor of a cable tool rig where visitors sit or the crew rests between chores.

**lead tong**: See backup tong.

**lease hound**: Oil lease locator and buyer, otherwise called a land man.

**lease house**: A single house situated on a lease where the pumper or other employee of a small independent oil company lives.

**the line**: The criminal group that controlled illegal alcohol sales during oil booms.

**location**: The place where an oil activity takes place, as in a drilling location. Also used to denote being present for activities of any kind, as in "He was on location when so and so occurred."

**lost circulation**: Losing the circulation of drilling fluid in a rotary drilling operation due to drilling into a cavity or some other similar situation.

**magazine**: Storage facility for the nitroglycerin used to shoot wells.

**major**: One of the large corporate oil companies that control all aspects of the oil business from exploration and production through marketing.

**make a boom**: To work during a particular oil boom, as in, "I made the Ranger boom back during 1918."

**marmon board**: Board about five feet long pulled by a team of draft animals to scrape dirt atop a pipeline after it was laid in its ditch.

**monkey board**: The platform upon which the derrick man works on a rotary rig. Also called a treble or thribble board when pulling stands of three joints (approximately 60 feet of pipe) and a forble board when pulling stands of four joints (approximately 80 feet of pipe).

**motor man**: Worker on a drilling rig responsible for tending to the engines. On the earlier steam-powered drilling rigs his job was referred to as the boiler man.

**mud pit**: Earthen pit (in the 1950s it became a portable metal receptacle instead of earthen) adjacent to a rotary rig location where the drilling mud is mixed and from which the mud is circulated into the well drilling operation.

**mule's foot**: A handheld chisel with an oval cutting edge instead of a straight cutting edge used by tank builders to cut bolt holes in metal.

**nipple chaser**: Worker who locates supplies for others. Derives from looking for nipples or short pieces of pipe used in connection work.

**oil patch**: The oil field as termed by the workers in the industry. The term has been used since the first days of the industry in Texas. It harks back to the rural roots of those early-day workers who traded in the cotton patch for the oil patch as a workplace.

**on the line**: Tank being emptied into the pipeline, as in "I'm putting this tank on the line."

**overshot**: Fishing tool device designed to go over the object lost in the hole in order to retrieve it.

**pad**: The raised dirt portion of a location upon which a drilling rig, tank battery, or other type of oilfield installation is constructed.

**pay sand**: Portion of an oil or gas formation where commercial production is found. Often simply referred to as the pay.

**pig**: Device placed inside a pipeline and forced through its length by pump pressure in order to clean the inside of the pipe.

**pipeliner**: Worker who lays pipelines.

**poor boy camp**: Oil company camps reserved for the hourly employees' residences. Also in reference to the residences of those oil company employees unable to qualify for company quarters who live in a cluster near the formal company camp.

**poor boy outfit**: Those companies, usually associated with the first thirty or forty years of the industry in Texas, who drilled wells and operated their leases with very few resources in the hope of striking it rich.

**pull it green**: Pulling a drill bit out of the hole before it is worn enough for replacement.

**pumper**: Oil company employee responsible for operating and maintaining the oil well pumps on a lease.

**rabbit**: A small plug run through a pipeline to check for obstructions.

**rag line**: Manila rope used as a drilling line on cable tool rigs prior to the introduction of steel cable around the beginning of the twentieth century.

**rat hole**: A 30- to 35-foot cased hole that projects above the drilling floor on a rotary rig where the kelly is stored while the pipe hoisting operation is in progress.

**returns**: See cuttings.

**rig ax**: Heavy duty hatchet with a long handle whose head is composed of blade on one side and a hammerhead on the other, used by rig builders to construct wooden derricks.

**rig builder**: Worker who builds and dismantles the derricks for drilling rigs. On cable tool rigs he also assembles the various wooden parts of the drilling equipment (bull wheels, calf wheels, band wheels, etc.).

**rigging down**: Dismantling a drilling rig and auxiliary equipment after finishing a well.

**rigging up**: Setting up a drilling rig and auxiliary equipment in preparation for drilling a well.

**rig irons**: Various iron parts of the power transmission section of a cable tool drilling rig.

**rockhound**: Name for a geologist.

**roughneck**: Term used collectively for the four workers supervised by the driller on a rotary rig whose specific names are motor man, derrick man, and two floor hands. Many prefer to refer to only the floor hands as roughnecks.

**round trip**: See trip.

**rotary drilling rig**: Type of drilling rig utilizing an augurlike motion with a bit attached to drill pipe to drill a well. It was introduced into Texas in 1895 and by 1930 became the primary drilling method used in the oilfield.

**rotary table**: The revolving table on a rotary drilling rig that causes the drill pipe to turn.

**roustabout**: Worker who does general maintenance work around oil leases. Roustabout crews are also referred to as connection crews or bull gangs.

**samson post**: The vertical beam that supports the walking beam on a cable tool rig.

**sand line**: The cable on a cable tool drilling rig that raises and lowers the bailer.

**scissors**: See tongs

**scout**: Person who investigates potential oil finds in a given area so he can lease the surrounding property ahead of other investigators.

**setter**: The boss of a tank-building crew. The name derives from the act of setting or building a tank.

**shooter**: The person who explodes nitroglycerin in a well in order to fracture the oil-bearing strata and allow more oil to flow into the well bore.

**shotgun house**: Type of small house common in boomtowns designed with either two or three rooms built in a line one behind the other. It purportedly gets its name from the fact that if a shotgun is fired through the front door it would exit out the back door without striking any walls. The shotgun house is also called the boxcar house due to its dimensions resembling a railroad boxcar.

**shut in**: The practice of closing off the flow of a producing oil well.

**sills**: The large support beams placed beneath wooden drilling rigs.

**skidding**: The act of moving a standard drilling derrick in toto.

**slush pit**: Earthen pit adjacent to a tank battery where salt water is drained into from the tanks, also a pit adjacent to a drilling rig where the drilling mud goes upon exiting the well bore.

**snapper**: Pipeliner who has one of the less arduous jobs referred to as snaps.

**spinning chain**: Length of chain used to wrap around drill pipe by a roughneck in order to spin the pipe up snugly before it is finally tightened by the use of tongs.

**spud wrench**: A hand tool composed of a wrench on one end and an alignment spike on the other used by rig builders to construct steel derricks.

**spudding in**: The process of drilling the first several feet of an oil well before changing over to the standard drilling procedure. The term derives from the early cable tool practice of using a spudding process that involved operating the drilling cable from the derrick top instead of from the walking beam.

**squat**: The amount of compression, measured in inches, on a wooden drilling derrick during the drilling process.

**squib**: One or more sticks of dynamite with a fuse used to explode the nitroglycerin in the well-shooting procedure.

**stabber**: A pipeline worker who aligns the pipe so it can be either screwed or welded together.

**stacking a rig**: Putting a drilling rig in storage.

**stand**: Two or more joints of pipe removed from a well as one section.

**standard drilling rig**: The wooden and steel drilling rigs, both cable tool and rotary, that had to be assembled before use and disassembled for moving to a new location. They were replaced by the jackknife rigs in the 1950s.

**stave**: A side piece for either a wood or a metal tank.

**strapper**: Pipeline worker who measures or straps a tank in order to verify the volume of oil in the vessel.

**string of tools**: The equipment used to drill an oil well. It is usually associated with early cable tool rigs where the drilling string comprised the primary metal objects used for drilling a well. A typical statement using the term might be, "Joe Blow brought the first string of tools into this field, back about the time of World War I."

**swamper**: Worker who acts as assistant to an oilfield truck driver.

**swivel**: The connection between the kelly and the mud pumps on a rotary rig.

**tank builder**: Worker who builds oil storage tanks.

**tank farm**: Concentration of a large number of giant tanks to supply either a pipeline or a refinery. The term harks back to the rural origin of early-day oilfield workers.

**tankie**: A tank builder.

**t.d. (total depth)**: To reach the desired drilling depth for completing a well.

**teaming**: Work done by the teamsters in the days prior to the use of motor trucks. Also used in reference to dirt contracting work utilizing animal-drawn power.

**temper screw**: The device hanging from the walking beam on a cable tool rig to which the drilling line is attached.

**thief hatch**: Opening in the deck of a tank through which measurements are made and samples are drawn. Derived from the idea of stealing the oil by the oil sampler or thief.

**thribble board**: See monkey board.

**throwing the chain**: Process of throwing the spinning chain around the drill pipe in order to spin the pipe up snugly before final tightening.

**tongs**: Large pipe wrenches used to tighten and loosen pipe. Usually associated with pipeliners and drilling crews.

**tool dresser**: Assistant to the driller on a cable tool rig. The name derives from the practice of dressing or sharpening the cable tool bit, which was part of his responsibility.

**tool pusher**: Employee in overall charge of one or more rotary drilling rigs.

**torpedo**: Tin tube that is filled with nitroglycerin in order to shoot an oil well.

**tour**: Pronounced "tower," it is the shift worked by drilling crews. From the turn of the century until the 1930s it consisted of two twelve-hour shifts that operated from noon until midnight and from midnight until noon. In the 1930s the tour was changed to three eight-hour shifts.

**traveling blocks**: The moving pulley block on a rotary drilling rig that raises and lowers drill pipe and casing.

**treble board**: See monkey board.

**trip**: The process of removing the drill pipe from the hole on a rotary rig. The process of removing the pipe and replacing it is called a round trip.

**turning to the right**: Term used in the oilfield to denote accomplishing work. It derives from the clockwise turning of the drill pipe during the drilling process.

**twist off**: A term used by oilfield hands in reference to making some sort of mistake usually associated with getting intoxicated. It derives from the on-the-job mistake of twisting the drill pipe apart during the drilling operation.

**undershot**: Fishing tool designed to go inside a pipe lost downhole.

**upstream**: That portion of the oil industry that takes place prior to refining.

**walking beam**: The horizontal beam on a cable tool rig whose rocking motion provides the striking force of the drill bit.
**water table**: The platform atop a drilling rig derrick on which the crown block rests.
**weight indicator**: The device used on a rotary drilling rig that indicates the total weight pressing down on the drill bit.
**wildcat**: A well drilled some distance from proven production with little chance of making a successful well.
**wildcatter**: Someone who drills wildcat wells and has the reputation of taking great chances.
**worm**: See boll weevil.

# Notes

Due to the large number of oral history interviews and the heavy use of the periodical *Oil Weekly*, the following abbreviations are used in the notes and in the bibliography:

CAH—"Pioneers in Texas Oil Collection" at the Center for American History, University of Texas, Austin, Texas
PBPM—Archives, Permian Basin Petroleum Museum, Midland, Texas
PPHM—Archives, Panhandle-Plains Historical Museum, Canyon, Texas
SFA—East Texas Research Center, Stephen F. Austin University, Nacogdoches, Texas
SWC—Southwest Collection, Texas Tech University, Lubbock, Texas
*OW*—*Oil Weekly*

## CHAPTER 1

1. Edgar Wesley Owen, *Trek of the Oil Finders: A History of Exploration for Oil* (Tulsa: American Association of Petroleum Geologists, 1975), 192–96; John O. King, *Joseph Stephen Cullinan: A Study of Leadership in the Texas Petroleum Industry, 1897–1937* (Nashville: Vanderbilt University Press, 1970), 88–90; Diana Davids Hinton and Roger Olien, *Oil in Texas: The Gusher Age, 1895–1945* (Austin: University of Texas Press, 2002), 4–7; Carl Coke Rister, *Oil: Titan of the Southwest* (Norman: University of Oklahoma Press, 1949), 43–49.

2. Robert W. McDaniel and Henry C. Dethloff, *Pattillo Higgins and the Search for Texas Oil* (College Station: Texas A&M University Press, 1989), 43–53.

3. Curt Hamill, interviewed by William A. Owens, July 17, 1952 (CAH); Allen W. Hamill, interviewed by William A. Owens, September 2, 1952 (CAH).

4. Curt Hamill, interviewed by William A. Owens, July 17, 1952 (CAH); Allen W. Hamill, interviewed by William A. Owens, September 2, 1952 (CAH); Curtis G. Hamill, *We Drilled Spindletop* (Houston: privately published, 1957), 4.

5. Curt Hamill, interviewed by William A. Owens, July 17, 1952 (CAH).

6. Bruce Ashburn, interviewed by Paul Patterson, April 29, 1968 (SWC). Ashburn gives a description of common rig-building practices in West Virginia ca. 1912 that is almost identical to the method the Hamills used at Spindletop. Curt Hamill, interviewed by William A. Owens, July 17, 1952 (CAH); Allen W. Hamill, interviewed by William A. Owens, September 2, 1952 (CAH). In his interview Curt Hamill gives the height of the derrick as eighty-four feet, but all other accounts give it as sixty feet. A close inspection

of photographs of the derrick also supports the sixty-foot height as being correct. In later years eighty-four feet was a standard height and perhaps Mr. Hamill confused the numbers fifty years after the event.

7. Curt Hamill, interviewed by William A. Owens, July 17, 1952 (CAH); Allen W. Hamill, interviewed by William A. Owens, September 2, 1952 (CAH); Hamill, *We Drilled Spindletop,* 22–25. Most of their contemporaries give the Hamill brothers credit for developing drilling mud. However, the use of mud-impregnated drilling fluid to seal off problematic porous formations had been known and used as early as 1887, although it was not a widely accepted practice. J. E. Brantly, *History of Oil Well Drilling* (Houston: Gulf Publishing Company, 1971), 1122–24.

8. Curt Hamill, interviewed by William A. Owens, July 17, 1952 (CAH); Allen W. Hamill, interviewed by William A. Owens, September 2, 1952 (CAH); Judith Walker Linsley, Ellen Walker Rienstra, and Jo Ann Stiles, *Giant under the Hill* (Austin: Texas State Historical Association, 2002), 89–110, gives the most detailed account of the drilling of the well available in print; Paul N. Spellman, *Spindletop Boom Days* (College Station: Texas A&M University Press, 2001), 32–46, also gives an excellent account of the actual drilling of the well.

9. *Beaumont Enterprise,* January 12, 14, 15, 18, 1901; *Beaumont Journal,* January 22, 1901; *Houston Post,* April 28, 1901.

10. James A. Clark and Michel T. Halbouty, *Spindletop* (New York: Random House, 1952), 90; Linsley, Rienstra, and Stiles, *Giant under the Hill,* 134, 166, 190–92; Spellman, *Spindletop Boom Days,* 62–63; Rister, *Oil: Titan of the Southwest,* 61–66; The only reliable population number for Beaumont during this era is the 1900 census that gives its population as approximately 9,500. All population figures relating to oil boomtowns should be viewed with great skepticism. The shock of seeing the tremendous numbers of people flocking to Beaumont upon the discovery of oil established a pattern for succeeding oil booms in which wildly exaggerated numbers in the tens of thousands of people are credited with populating those entities when in fact there are no solid records available to substantiate those claims. For a good case study of this phenomenon, see Roger M. Olien and Diana Davids Olien, *Oil Booms: Social Change in Five Texas Towns* (Lincoln: University of Nebraska Press, 1982), 21–40.

11. Sam Webb, interviewed by William A. Owens, September 1952 (CAH); Benjamin "Bud" Coyle, interviewed by William A. Owens, July 28, 1953 (CAH).

12. Frank Redman, interviewed by William A. Owens, July 20, 1953 (CAH).

13. Claude Deer, interviewed by William A. Owens, July 7, 1952 (CAH); Harry R. Paramore, interviewed by William A. Owens, July 2, 1953 (CAH); Judge Edgar Eggleston Townes, interviewed by William A. Owens, June 21, 1952 (CAH).

14. Frank Redman, interviewed by William A. Owens, July 20, 1953 (CAH); Sam W. Webb, interviewed by William A. Owens, September 1952 (CAH); Claude Deer, interviewed by William A. Owens, July 7, 1952 (CAH); Benjamin "Bud" Coyle, interviewed

by William A. Owens, July 28, 1953 (CAH); R. R. Hobson, interviewed by William A. Owens, May 15, 1956 (CAH).

15. Clark and Halbouty, *Spindletop*, 98–99; Judge Edgar Eggleston Townes, interviewed by William A. Owens, June 21, 1952 (CAH).

16. Dr. H. C. Sloop, interviewed by William A. Owens, March 25, 1955 (CAH).

17. Frank Redman, interviewed by William A. Owens, July 20, 1953 (CAH).

18. Dr. H. C. Sloop, interviewed by William A. Owens, March 25, 1955 (CAH).

19. Judith Walker Linsley and Ellen Walker Rienstra, *Beaumont: A Chronicle of Promise* (Woodland Hills, Calif.: Windsor Publications, 1982), 83; Claude L. Witherspoon, interviewed by Mody Boatright, June 27, 1953 (CAH); Benjamin "Bud" Coyle, interviewed by William A. Owens, July 28, 1953 (CAH).

20. William L. Armstrong, interviewed by William A. Owens, July 24, 1953 (CAH); Dr. D. W. Davis, interviewed by William A. Owens, July 18, 1953 (CAH).

21. James William Kinnear, interviewed by William A. Owens, July 16, 1953 (CAH).

22. William L. Armstrong, interviewed by William A. Owens, July 24, 1953 (CAH).

23. Hinton and Olien, *Oil in Texas*, 42; Benjamin "Bud" Coyle, interviewed by William A. Owens, July 28, 1953 (CAH); James William Kinnear, interviewed by William A. Owens, July 16, 1953 (CAH); Dr. D. W. Davis, interviewed by William A. Owens, July 18, 1953 (CAH); H. P. Nichols, interviewed by Mody Boatright, October 11, 1952 (CAH); Mr. and Mrs. Sam W. Webb, interviewed by William A. Owens, September 1952 (CAH).

24. Benjamin "Bud" Coyle, interviewed by William A. Owens, July 28, 1953 (CAH); Hardeman Roberts, interviewed by William A. Owens, April 26, 1956 (CAH); Mr. and Mrs. Sam W. Webb, interviewed by William A. Owens, September 1952 (CAH); Rister, *Oil: Titan of the Southwest*, 76.

25. Hardeman Roberts, interviewed by William A. Owens, April 26, 1956 (CAH); M. A. "Curley" Johnson, interviewed by Mody Boatright, August 27, 1952 (CAH); Frank Hamilton, interviewed by Mody Boatright, September 29, 1952 (CAH); W. H. "Bill" Bryant, interviewed by William A. Owens, July 29, 1952 (CAH).

26. Benjamin A. "Bud" Coyle, interviewed by William A. Owens, July 28, 1953 (CAH); W. H. "Bill" Bryant, interviewed by William A. Owens, July 29, 1952 (CAH).

27. Mr. and Mrs. Sam W. Webb, interviewed by William A. Owens, September, 1952 (CAH); Early C. Deane, interviewed by William A. Owens, July 10, 1953 (CAH).

28. William Edward Cotton, interviewed by William A. Owens, May 23, 1956 (CAH).

29. F. G. Swanson, interviewed by R. M. Hayes, February 20, 1955 (CAH); C. C. McClelland, interviewed by Mody Boatright, September 9, 1952 (CAH); Mr. and Mrs. Sam Webb, interviewed by William A. Owens, September 1952 (CAH); Mr. And Mrs. V. B. Daniels, interviewed by William A. Owens, July 3, 1952 (CAH).

30. William Warren Sterling, *Trails and Trials of a Texas Ranger* (Norman: University of Oklahoma Press, 1968), 326–31.

## CHAPTER 2

1. Brantly, *History of Oil Well Drilling*, 10–12, 546; M. A. "Curly" Johnson, interviewed by Mody Boatright, August 27, 1952 (CAH).

2. Curt G. Hamill, interviewed by William A. Owens, July 16, 1952 (CAH); Frank Redman, interviewed by William A. Owens, July 20, 1953 (CAH); William Joseph Philp, interviewed by William A. Owens, July 17, 1953 (CAH); Fred Jennings, interviewed by William A. Owens, June 19, 1952 (CAH). I have chosen to call all drilling crew members except the driller *roughnecks*, although some choose to call only the two floor hands roughnecks.

3. H. P. Nichols, interviewed by Mody Boatright, October 11, 1952 (CAH).

4. Sam W. Webb, interviewed by William A. Owens, September 1952 (CAH).

5. W. H. "Bill" Bryant, interviewed by William A. Owens, July 29, 1952 (CAH); Benjamin "Bud" Coyle, interviewed by William A. Owens, July 28, 1953 (CAH); Sam W. Webb, interviewed by William A. Owens, September 1952 (CAH); James William Kinnear, interviewed by William A. Owens, July 16, 1953 (CAH); E. M. Friend, interviewed by Mody Boatright, September 4, 1953 (CAH); J. A. Rush, interviewed by Mody Boatright, September 11, 1956 (CAH).

6. Benjamin "Bud" Coyle, interviewed by William A. Owens, July 29, 1952 (CAH).

7. James William Kinnear, interviewed by William A. Owens, July 16, 1953 (CAH); W. H. "Bill" Bryant, interviewed by William A. Owens, July 29, 1952 (CAH); Frank Redman, interviewed by William A. Owens, July 20, 1953 (CAH); E. M. Friend, interviewed by Mody Boatright, September 4, 1953 (CAH); Benjamin "Bud" Coyle, interviewed by William A. Owens, July 28, 1953 (CAH).

8. Curt G. Hamill, interviewed by William A. Owens, July 17, 1952 (CAH).

9. Dr. D. W. Davis, interviewed by William A. Owens, July 17, 1952 (CAH); William Joseph Philp, interviewed by William A. Owens, July 17, 1953 (CAH); Hardeman Roberts, interviewed by William A. Owens, April 26, 1956 (CAH); Sam W. Webb, interviewed by William A. Owens, September 1952 (CAH); Frank Hamilton, interviewed by William A. Owens, September 29, 1952 (CAH).

10. Curt G. Hamill, interviewed by William A. Owens, July 27, 1952 (CAH).

11. Benjamin "Bud" Coyle, interviewed by William A. Owens, July 28, 1953 (CAH); Claude Deer, interviewed by William A. Owens, July 7, 1953 (CAH); James Donohoe, interviewed by William A. Owens, August 1, 1952 (CAH); W. H. "Bill" Bryant, interviewed by William A. Owens, July 29, 1952 (CAH). As a result of this fire and the ongoing danger of similar situations a safety committee made up of representatives from the various operators at Spindletop was established in 1902, but fire remained a source of significant danger for years to come. See Linsley, Rienstra, and Stiles, *Giant under the Hill*, 163–66.

12. W. H. "Bill" Bryant, interviewed by William A. Owens, July 29, 1952 (CAH).

13. Carl F. Mirus, interviewed by William A. Owens, April 4, 1956 (CAH); Bert Stiv-

ers, interviewed by William A. Owens, August 26, 1953 (CAH); W. H. "Bill" Bryant, interviewed by William A. Owens, July 29, 1952 (CAH).

14. W. H. "Bill" Bryant, interviewed by William A. Owens, July 29, 1952 (CAH); Benjamin "Bud" Coyle, interviewed by William A. Owens, July 28, 1953 (CAH).

15. John D. Alexander, interviewed by Richard Mason, October 20, 1982 (SWC).

16. Walter Cline, interviewed by Mody Boatright, August 18, 1952 (CAH); Claude Deer, interviewed by William A. Owens, July 7, 1952 (CAH); W. H. "Bill" Bryant, interviewed by William A. Owens, July 29, 1952 (CAH); W. M. Hudson, interviewed by W. M. Hudson Jr., September 18, 1952 (CAH); F. G. Swanson, interviewed by R. M. Hayes, February 20, 1955 (CAH).

17. Benjamin "Bud" Coyle, interviewed by William A. Owens, July 28, 1953 (CAH); H. P. Nichols, interviewed by Mody Boatright, October 11, 1952 (CAH); William Joseph Philp, interviewed by William A. Owens, July 17, 1953 (CAH); Sam W. Webb, interviewed by William A. Owens, September 1952 (CAH); Harry R. Paramore, interviewed by William A. Owens, July 2, 1953 (CAH); Frank Redman, interviewed by William A. Owens, July 20, 1953 (CAH); Fred Jennings, interviewed by William A. Owens, July 19, 1952 (CAH); E. M. Friend, interviewed by Mody Boatright, September 14, 1953 (CAH); A. R. Dillard, interviewed by William A. Owens, September 5, 1953 (CAH).

18. J. A. Rush, interviewed by Mody Boatright, September 11, 1953 (CAH); Brantly, *History of Oil Well Drilling,* 1211–19, 1174–76.

19. John H. Wynn, interviewed by William A. Owens, July 4, 1955 (CAH).

20. John S. Wynn, interviewed by William A. Owens, July 4, 1955 (CAH); J. A. Rush, interviewed by Mody Boatright, September 11, 1953 (CAH); Brantly, *History of Oil Well Drilling,* 1211–19, 1174–76.

21. The various prejudices of the advantages of cable tool over rotary and the relative merits of those beliefs are examined in Stanley Gill, "A Study of Rotary Drilling Mud," in *OW,* June 13, 1932, 24–36.

22. Dan Lively, interviewed by Paul Patterson, May 23, 1968 (SWC).

23. Claude Deer, interviewed by William A. Owens, July 7, 1952 (CAH); James Donohoe, interviewed by William A. Owens, August 1, 1952 (CAH); J. A. Rush, interviewed by Mody Boatright, September 11, 1956 (CAH); Benjamin "Bud" Coyle, interviewed by William A. Owens, July 29, 1953 (CAH).

24. Mody Boatright, *Folklore of the Oil Industry* (Dallas: Southern Methodist University Press, 1963), 199.

25. Dan Lively, interviewed by Paul Patterson, May 23, 1968 (SWC); Ed P. Mattson, interviewed by Mody Boatright, June 19, 1953 (CAH); Walter Cline, interviewed by Mody Boatright, August 18, 1952 (CAH); Landon Haynes, interviewed by Robert O. Stephens, July 31, 1959 (CAH); H. A. Rathke, interviewed by Mody Boatright, September 13, 1953 (CAH); E. M. Friend, interviewed by Mody Boatright, September 4, 1953 (CAH); A. R. Dillard, interviewed by Mody Boatright, September 5, 1953 (CAH).

26. Gerald Lynch, *Roughnecks, Drillers, and Tool Pushers* (Austin: University of Texas Press, 1987), 17–28.

## CHAPTER 3

1. Olien and Olien, *Oil Booms: Social Change in Five Texas Towns,* 6–11.

2. Dr. H. C. Sloop, interviewed by William A. Owens, March 25, 1955 (CAH).

3. Curt G. Hamill, interviewed by William A. Owens, August 17, 1952 (CAH); Plummer M. Barfield, interviewed by William A. Owens, August 1, 1952 (CAH); William Edward Cotton, interviewed by William A. Owens, May 23, 1956 (CAH).

4. Landon Haynes Cullum, interviewed by Robert O. Stephens, January 3, 1953 (CAH).

5. Plummer M. Barfield, interviewed by William A. Owens, August 1, 1952 (CAH); William Edward Cotton, interviewed by William A. Owens, May 23, 1956 (CAH); James Donohoe, interviewed by William A. Owens, August 1, 1952 (CAH); W. H. "Bill" Bryant, interviewed by William A. Owens, July 29, 1952 (CAH); F. G. Swanson, interviewed by R. M. Hayes, February 20, 1955 (CAH).

6. James William Kinnear, interviewed by William A. Owens, July 24, 1953 (CAH).

7. Plummer M. Barfield, interviewed by William A. Owens, August 1, 1952 (CAH).

8. Curt G. Hamill, interviewed by William A. Owens, August 17, 1952 (CAH); Allen W. Hamill, interviewed by William A. Owens, September 2, 1952 (CAH).

9. Brantly, *History of Oil Well Drilling,* 343–46; Louis C. Sands, "Oil-Field Development and Petroleum Production," in *Handbook of the Petroleum Industry,* ed. David T. Day (New York: John Wiley and Sons, 1923), 204–206; Oil Well Supply Company, *General Catalogue and Price List* (New York: Oil Well Supply Company, 1916), 4–7.

10. Brantly, *History of Oil Well Drilling,* 343–46.

11. Claude Deer, interviewed by William A. Owens, July 7, 1952 (CAH); Frank Redman, interviewed by William A. Owens, July 20, 1953 (CAH); L. D. Winfrey, interviewed by Mrs. Robert A. Montgomery, August 2, 1952 (CAH); W. H. Hudson, interviewed by W. H. Hudson Jr., September 18, 1952 (CAH); Frank Dunn, interviewed by William A. Owens, March 23, 1955 (CAH).

12. F. G. Swanson, interviewed by R. M. Hayes, February 20, 1955 (CAH); Charlie Storms, interviewed by Ned Dewitt, August 16, 1939, in Paul Lambert and Kenny Franks, eds., *Voices from the Oil Fields* (Norman: University of Oklahoma Press, 1984), 22–38.

13. "Skeet" Wagner, interviewed by Bobby Weaver, spring 1983 (author's collection).

14. W. H. "Bill" Bryant, interviewed by William A. Owens, July 29, 1952 (CAH). The term *pusher* that Mr. Bryant uses in this quote refers to the "tool pusher" who was the man in overall charge of the rig or possibly several rigs. Although the drillers were in charge of their drilling crew during normal shift operations, each driller was responsible to the tool pusher.

15. Frank Hamilton, interviewed by Mody Boatright, September 29, 1952 (CAH); H. A. Rathke, interviewed by Mody Boatright, September 13, 1953 (CAH).

16. For an excellent overview of the development of oil well shooting prior to oil development in Texas, see Sandy Gow, *Roughnecks, Rock Bits and Rigs: The Evolution of Oil Well Drilling Technology in Alberta, 1883–1970* (Calgary, Alberta: University of Calgary Press, 2005), 125–28.

17. Ed P. Matteson, interviewed by Mody Boatright, June 19, 1953 (CAH).

18. Frank Dunn, interviewed by William A. Owens, July 29, 1952 (CAH).

19. Max W. Ball, *This Fascinating Oil Business* (New York: Bobbs-Merrill, 1940), 110–12; Brantly, *History of Oil Well Drilling*, 1251–70; Day, ed., *Handbook of the Petroleum Industry*, 219–20.

20. Charles W. Randolph, interviewed by William A. Owens, spring 1955 (CAH).

21. Curt Hamill, interviewed by William A. Owens, July 17, 1953 (CAH).

22. James William Kinnear, interviewed by William A. Owens, July 24, 1953 (CAH); Frank Dunn, interviewed by William A. Owens, February 23, 1955 (CAH).

23. Frank Dunn, interviewed by William A. Owens, February 23, 1955 (CAH); *Gulf Coast Oil News*, March 17, 1917, 7; July 21, 1917, 17–21; *OW*, March 11, 1919, 32.

24. Curt G. Hamill, interviewed by William A. Owens, August 17, 1952 (CAH); Claude L. Witherspoon, interviewed by Mody Boatright, June 27, 1953 (CAH); Rister, *Oil: Titan of the Southwest*, 62–63.

25. Alexander Balfour Patterson, interviewed by William A. Owens, August 4, 1953 (CAH); Frank Dunn, interviewed by William A. Owens, March 23, 1955 (CAH); Benjamin "Bud" Coyle, interviewed by William A. Owens, July 28, 1956 (CAH).

26. *Gulf Coast Oil News*, September 22, 1917, 9; December 23, 1917, 33; February 23, 1918, 33; *OW*, August 24, 1918, 13; February 26, 1921, 33; April 29, 1922, 1; October 7, 1922, 1.

27. John L. Loos, *Oil on Stream! A History of Interstate Oil Pipe Line Company, 1909–1959* (Baton Rouge: Louisiana State University Press, 1959), 26; Lambert and Franks, eds., *Voices from the Oil Fields*, 198.

28. Lambert and Franks, eds., *Voices from the Oil Fields*, 203.

29. Burt E. Hull, interviewed by William A. Owens, August 24, 1953 (CAH); Ralph B. McLaighlin, interviewed by William A. Owens, May 28, 1956 (CAH); Loos, *Oil on Stream*, 26.

30. Burt E. Hull, interviewed by William A. Owens, August 24, 1953 (CAH).

31. Burt E. Hull, interviewed by William A. Owens, August 24, 1953 (CAH); Loos, *Oil on Stream*, 12, 26.

32. Burt E. Hull, interviewed by William A. Owens, August 24, 1953 (CAH); Loos, *Oil on Stream*, 14–17; Lambert and Franks, eds., *Voices from the Oil Fields*, 202–203; Ball, *This Fascinating Oil Business*, 179–80.

33. Burt E. Hull, interviewed by William A. Owens, August 24, 1953 (CAH); Loos, *Oil on Stream*, 17.

34. Burt E. Hull, interviewed by William A. Owens, August 24, 1953 (CAH); Loos, *Oil on Stream*, 19–23.

35. Burt E. Hull, interviewed by William A. Owens, August 24, 1953 (CAH); Ralph B. McLaughlin, interviewed by William A. Owens, May 28, 1956 (CAH).

36. Loos, *Oil on Stream,* 19, 20–23.

## CHAPTER 4

1. H. P. Nichols, interviewed by Mody Boatright, October 11, 1952 (CAH); Rister, *Oil: Titan of the Southwest,* 108–109.

2. Claude L. Witherspoon, interviewed by Mody Boatright, June 27, 1953 (CAH).

3. "Tom Waggoner Talks of Olden Days," *Oil and Gas Journal* (April 23, 1920): 64.

4. *Daily Times,* Wichita Falls, Texas, April 1, 1936.

5. A. R. Dillard, interviewed by Mody Boatright, September 5, 1953 (CAH). The situation of wooden derricks being toppled by high winds in the North Texas area and indeed across the oilfields of the state was a common occurrence as evidenced by numerous articles in industry publications such as those appearing in *OW,* May 15, 1920, 34 and June 26, 1920, 36.

6. Brantly, *History of Oil Well Drilling,* 13.

7. *OW,* November 12, 1918, 4, 22; April 26, 1919, 52.

8. Carl F. Mirus, interviewed by William A. Owens, April 4, 1956 (CAH).

9. Hinton and Olien, *Oil in Texas,* 78–79; Rister, *Oil: Titan of the Southwest,* 113–15.

10. Hinton and Olien, *Oil in Texas,* 83–84.

11. *OW,* September 23, 1918, 22; September 28, 1918, 9, 12; October 30, 1918, 28; December 17, 1918, 23.

12. Walter Cline, interviewed by Mody Boatright, August 18, 1952 (CAH); *Weekly Times,* Wichita Falls, Tex., August 2, 1918.

13. Thornton Lomax, interviewed by Richard Mason, January 15, 1982 (SWC).

14. Landon Haynes Cullum, interviewed by Robert O. Stephens, January 31, 1959 (CAH).

15. Thornton Lomax, interviewed by Richard Mason, January 15, 1982 (SWC); J. T. "Cotton" Young, interviewed by Mody Boatright, August 15, 1952 (CAH); John F. Donohoe, interviewed by Mody Boatright, September 4, 1953 (CAH); John D. Alexander, interviewed by Richard Mason, October 20, 1982 (SWC); Burk Paschall, interviewed by Mody Boatright, July 30, 1952 (CAH); J. R. Webb, interviewed by Mody Boatright, August 1, 1952 (CAH); *Oil City Derrick Statistical Abstract of Petroleum,* 4, Rister Collection, SWC; For a fascinating look at the activities of oil promotion during this era, see Roger M. and Diana Davids Olien, *Easy Money: Oil Promoters and Investors in the Jazz Age* (Chapel Hill: University of North Carolina Press, 1990).

16. Alexander Balfour Patterson, interviewed by William A. Owens, August 4, 1953 (CAH).

17. *OW,* October 5, 1918, 8; February 18, 1919, 32; October 18, 1919, 208; R. L. Dudley, "The Oil Worker Cared For," *OW,* October 23, 1920, 115–20.

18. Wallace Pratt, interviewed by Richard Mason, January 1, 1982 (SWC); Alexander

Balfour Patterson, interviewed by William A. Owens, August 4, 1953 (CAH); John D. Alexander, interviewed by Richard Mason, October 20, 1982 (SWC).

19. Jack Knight, interviewed by Mody Boatright, September 1952 (CAH); Alexander Balfour Patterson, interviewed by William A. Owens, August 4, 1953 (CAH); Wallace Cox, interviewed by Richard Mason, January 1, 1982 (SWC).

20. Charlie Davis, interviewed by Richard Mason, February 12, 1981 (SWC); Lloyd Bruce, interviewed by Richard Mason, January 15, 1981 (SWC); C. C. McClelland, interviewed by Mody Boatright, September 9, 1952 (CAH).

21. Rister, *Oil: Titan of the Southwest,* 143; *OW,* October 18, 1919, 83–86.

22. Ed P. Matteson, interviewed by Mody Boatright, June 19, 1953 (CAH); Charlie Davis, interviewed by Richard Mason, February 12, 1981 (SWC); Lloyd E. Bruce, interviewed by Richard Mason, January 15, 1981 (SWC); E. L. Lantron, interviewed by Mody Boatright, September 8, 1952 (CAH).

23. Jack Knight, interviewed by Mody Boatright, September, 1952 (CAH).

24. Frank Champion, interviewed by Ruth Hosey, n.d. (SWC); Grady Triplett, "Oil Field Thievery Is a Fine Art," *OW,* October 18, 1919, 237–40.

25. Lloyd E. Bruce, interviewed by Richie Cravens, November 26, 1976 (SWC); Lloyd E. Bruce, interviewed by Richard Mason, January 15, 1981 (SWC); Charlie Davis, interviewed by Richard Mason, February 12, 1981 (SWC); Rister, *Oil: Titan of the Southwest,* 145–46; Hinton and Olien, *Oil in Texas,* 79.

26. Lloyd E. Bruce, interviewed by Richie Cravens, November 26, 1976 (SWC); Lloyd E. Bruce, interviewed by Richard Mason, January 15, 1981 (SWC).

27. Mrs. M. R. Hamrick and Mrs. F. E. Langston, interviewed by Charles Townsend, April 7, 1967 (SWC).

28. Carl F. Mirus, interviewed by William A. Owens, April 4, 1956 (CAH).

29. A. J. Thaman, interviewed by Mrs. Maude Ross, January 29, 1960 (CAH).

30. *OW,* February 22, 1919, 18.

31. Lloyd E. Bruce, interviewed by Richie Cravens, November 26, 1976 (SWC); Lloyd E Bruce, interviewed by Richard Mason, January 15, 1981 (SWC); Mrs. M. R. Hamrick and Mrs. F. E. Langston, interviewed by Charles Townsend, April 4, 1967 (SWC).

32. Charles Laughlin, interviewed by Richard Mason, September 18, 1981 (SWC); Bert C. Bloodworth, interviewed by Bobby Weaver, January 17, 1978 (SWC); Charlie Davis, interviewed by Richard Mason, February 12, 1981 (SWC); Mrs. M. R. Hamrick and Mrs. F. E. Langston, interviewed by Charles Townsend, April 7, 1967 (SWC); John D. Alexander, interviewed by Richard Mason, October 20, 1982 (SWC); Bruce Ashburn, interviewed by Paul Patterson, April 29, 1968 (SWC); Hinton and Olien, *Oil in Texas,* 82–83.

33. Lloyd E. Bruce, interviewed by Richie Cravens, November 26, 1976 (SWC).

34. Lloyd E. Bruce, interviewed by Richard Mason, January 15, 1981 (SWC); Bert C. Bloodworth, interviewed by Bobby Weaver, January 17, 1978 (SWC); John D. Alexander, interviewed by Richard Mason, October 20, 1982 (SWC); for a good overview of the sensationalist literature related to lawlessness in Texas oil boomtowns and how it has

colored the popular perception of the character of oilfield workers, see Olien and Olien, *Oil Booms: Social Change in Five Texas Towns,* 209–10.

35. Wallace Pratt, interviewed by E. DeGolyer, March 2, 1945 (DeGolyer Papers, DeGolyer Library, Southern Methodist University, Dallas, Tex.); Mrs. M. R. Hamrick and Mrs. F. E. Langston, interviewed by Charles Townsend, April 7, 1967 (SWC); Charles Laughlin, interviewed by Richard Mason, September 28, 1981 (SWC); Rister, *Oil: Titan of the Southwest,* 157.

36. Landon Haynes Cullum, interviewed by Robert O. Stephens, January 3, 1959 (CAH); Walter Greenhaw, interviewed by Richard Mason, October 30, 1981 (SWC); Rister, *Oil: Titan of the Southwest,* 159–61; Hinton and Olien, *Oil in Texas,* 87.

37. Landon Haynes Cullum, interviewed by Robert O. Stephens, January 3, 1959 (CAH); Rister, *Oil: Titan of the Southwest,* 160–61; *Fort Worth Star Telegram,* September 3, 1918; *The Dublin (Texas) Progress,* September 6, 1918.

38. Landon Haynes Cullum, interviewed by Robert O. Stephens, January 3, 1959 (CAH).

39. Mr. and Mrs. (hereafter M&M) Guy Bosworth, interviewed by Paul Patterson, June 12, 1968 (SWC); C. C. McClelland, interviewed by Mody Boatright, September 9, 1952 (CAH); Robert H. Crawford, interviewed by Bobby Weaver, January 1, 1978 (SWC); Inez Heeter, interviewed by Richard Mason, October 30, 1981 (SWC); E. E. Brackens, interviewed by Samuel D. Myres, July 16, 1970 (PBPM); *Dublin (Texas) Progress,* August 29, 1919.

40. Walter Greenhaw, interviewed by Richard Mason, October 30, 1981 (SWC); Inez Heeter, interviewed by Richard Mason, October 30, 1981 (SWC); M&M Guy Boswell, interviewed by Paul Patterson, June 12, 1968 (SWC); Robert H. Crawford, interviewed by Bobby Weaver, January 1, 1978 (SWC).

41. *Dallas Morning News,* August 30 and October 9, 1920.

42. O. G. Lawson, interviewed by Mody Boatright, July 28, 1952 (CAH); Carl Angstadt, interviewed by Mody Boatright, August 4, 1952 (CAH); Ed P. Matteson, interviewed by Mody Boatright, June 19, 1953 (CAH).

43. O. G. Lawson, interviewed by Mody Boatright, July 28, 1952 (CAH); Carl Angstadt, interviewed by Mody Boatright, August 4, 1952 (CAH)

44. H. A. Rathke, interviewed by Mody Boatright, September 13, 1953 (CAH).

45. Ed P. Matteson, interviewed by Mody Boatright, June 19, 1953 (CAH).

46. Of the more than 350 oral history interviews used for this study at least fifty mentioned that much of the oilfield business done during that first twenty or thirty years of its existence in Texas was done on a handshake basis.

## CHAPTER 5

1. Esther Klinke, interviewed by Bobby Weaver, July 29, 1980 (PPHM); Henry Bellinghausen, interviewed by Bobby Weaver, July 29, 1980 (PPHM); Charles N. Gould, "How We Found the John Ray Dome at Amarillo," Box 22, Charles N. Gould Collection,

Western History Collection, University of Oklahoma, Norman; Charles N. Gould, "The Beginnings of the Panhandle Oil and Gas Field," *Panhandle-Plains Historical Review* 12 (1939): 55–63; Charles N. Gould, "Report on Geological Conditions Northeast of Amarillo, Texas," n.d., Panhandle-Plains Historical Museum, Canyon, Texas.

2. Gould, "Beginnings of the Panhandle Oil and Gas Field," 57.

3. Charles N. Gould, *Covered Wagon Geologist* (Norman: University of Oklahoma Press, 1959), 191–94; Gould, "Beginnings of the Panhandle Oil and Gas Field," 55–56; Wallace E. Pratt, "Oil and Gas in the Texas Panhandle" *Bulletin of the American Association of Petroleum Geologists* 7 (January–December, 1923): 237.

4. *Daily Panhandle,* (Amarillo, Texas), February 6, 1919; Thomas S. Cartwright, "History of Pioneer Natural Gas Company," *Panhandle Petroleum,* ed. Bobby D. Weaver (Canyon, Tex.: Panhandle-Plains Historical Society, 1982), 73–80; M. C. Nobles to Dr. Charles N. Gould, April 22, 1929, Rister Collection, SWC.

5. Eugene S. Blasdel to J. Evetts Haley, July 3, 1926, Panhandle-Plains Historical Museum; Gould, "Beginnings of the Panhandle Oil and Gas Field," 55–56.

6. Bureau of Economic Geology at the University of Texas, *Atlas of Major Texas Oil Reservoirs* (Austin: University of Texas Press, 1983), 135–37; Lawrence R. Hagy, "History and Development of the General Geology of the Panhandle Field of Texas," in Weaver, ed., *Panhandle Petroleum,* 3–9; N. D. Bartlett, "Discovery of the Panhandle Oil and Gas Field," in Weaver, ed., *Panhandle Petroleum,* 52.

7. "Pioneer Panhandle Oilman Dies," *Amarillo Globe,* August 27, 1945; "McIlroy Brothers Pioneers in Field," *Amarillo Sunday News,* n.d., Clippings File, Panhandle-Plains Historical Museum; Bartlett, "Discovery of the Panhandle Oil and Gas Field," 51.

8. Lawrence Hagy, interviewed by James Gamberton, August 8, 1978 (PPHM); M&M R. L. Johnson and Jewel Keith, interviewed by Mary E. Davidson, April 27, 1976 (PPHM); E. B. Garrett, "Oil Is King in the Texas Panhandle," *World's Work* (Kansas City: Kansas City Star, December, 1927), 170; Bartlett, "Discovery of the Panhandle Oil and Gas Field," 51.

9. M&M D. W. Page, interviewed by Mary E. Davidson, April 22, 1976 (PPHM); C. Don Hughes, interviewed by Amelia S. Nelson, November 4, 1979 (PPHM); M. T. Johnson Sr., interviewed by Marjorie L. Morris, June 27, 1972 (PPHM); P. J. R. McIntosh, "The Wonder Story of Texas Oil," *Texas Monthly* 3 (January–June, 1929), 208; *Oil World,* June 25, 1926, 42; *Amarillo Daily News,* March 6, 1927; March 8, 1927.

10. M&M R. L. Johnson and Jewel Keith, interviewed by Mary E. Davidson, April 27, 1976 (PPHM); Mrs. Claude Ruby, interviewed by Mary E. Davidson, November 7, 1975 (PPHM).

11. M&M D. W. Page, interviewed by Mary E. Davidson, April 27, 1976 (PPHM); Mrs. Aline Flaugher, interviewed by Mary E. Davidson, April 13, 1976 (PPHM); George Finger, interviewed by Mary E. Davidson, November 19, 1975 (PPHM); Mrs. Jimmie Cunningham, interviewed by Mary E. Davidson, February 19, 1976 (PPHM); Fritz Thompson, interviewed by Mary E. Davidson, April 13, 1976 (PPHM); David Warren, interviewed

by Mary E. Davidson, April 15, 1976 (PPHM); Mrs. Claude Ruby, interviewed by Mary E. Davidson, November 7, 1975 (PPHM); Gus Keith, interviewed by Bobby Weaver, March 4, 1982 (PPHM); M&M Burt Bryan, interviewed by S. D. Thompson, April 17, 1975 (PPHM); O. B. Hunt, interviewed by Dennis R. Boren, April 19, 1980 (PPHM).

12. *Dallas Morning News,* April 19, 1927; Borger *Daily Herald,* August 7, 1927; September 30, 1929; October 30, 1929; *Report of the Adjutant General of the State of Texas for the Fiscal Year, September 1, 1929 to August 31, 1930,* 51–52; John H. Jenkins and Gordon Frost, *I'm Frank Hamer* (Austin: State House Press, 1993), 144–50.

13. Mrs. Claude Ruby, interviewed by Mary E. Davidson, November 7, 1975 (PPHM); Gus Keith, interviewed by Bobby Weaver, March 4, 1982 (PPHM); M. T. Johnson, interviewed by Marjorie L. Morris, June 27, 1972 (PPHM); Lawrence Hagy, interviewed by James H. Gamberton, August 8, 1978 (PPHM); Jo Ann Rumpel (reminiscence), n.d. (PPHM); C. Don Hughes, interviewed by Bobby Weaver, July 23, 1982 (PPHM).

14. M&M R. L. Johnson and Jewel Keith, interviewed by Mary E. Davidson, April 27, 1976 (PPHM); Mrs. Claude Ruby, interviewed by Mary E. Davidson, November 7, 1975 (PPHM).

15. Gus Keith, interviewed by Bobby Weaver, March 4, 1982 (PPHM).

16. M&M D. W. Page, interviewed by Mary E. Davidson, April 22, 1976 (PPHM); Jack Knight, interviewed by Mary E. Davidson, April 13, 1976 (PPHM); Ware Harder, interviewed by Mary E. Davidson, April 8, 1976 (PPHM); Mrs. Aline Flaugher, interviewed by Mary E. Davidson, April 13, 1976 (PPHM); Cline Edwards, interviewed by Mary E. Davidson, April 15, 1976 (PPHM); Max Sherman, interviewed by Mary E. Davidson, April 27, 1976 (PPHM); Fritz Thompson, interviewed by Mary E. Davidson, April 13, 1976 (PPHM); M&M R. L. Johnson and Jewel Keith, interviewed by Mary E. Davidson, April 27, 1976 (PPHM); Mrs. Claude Ruby, interviewed by Mary E. Davidson, November 7, 1975 (PPHM); Tom Sappington and J. Keith, interviewed by Mary E. Davidson (PPHM); Jo Ann Rumpel (reminiscence), n.d. (PPHM); M&M Burt Bryan, interviewed by S. D. Thompson, April 7, 1975 (PPHM).

17. George Finger, interviewed by Mary E. Davidson, November 19, 1975 (PPHM); Sharon Chase, interviewed by Mary E. Davidson, May 10, 1976 (PPHM); Don Baker, interviewed by Mary E. Davidson, April 27, 1976 (PPHM); Fritz Thompson, interviewed by Mary E. Davidson, April 13, 1976 (PPHM); Mrs. Claude Ruby, interviewed by Mary E. Davidson, November 7, 1975 (PPHM).

18. Jack Knight, interviewed by Mary E. Davidson, April 13, 1976 (PPHM); M&M D. W. Page, interviewed by Mary E. Davidson, April 22, 1076 (PPHM); Ware Harder, interviewed by Mary E. Davidson, April 8, 1976 (PPHM); Horace W. Hickox, interviewed by Don Abbe, March 15, 1982 (PPHM).

19. Mrs. F. L. Cox and Mary Moser, interviewed by Mary E. Davidson, May 20, 1976 (PPHM); Frank Castleberry, interviewed by Mary E. Davidson, February 12, 1976 (PPHM); M&M R. L. Johnson and Jewel Keith, interviewed by Mary E. Davidson, April 27, 1976 (PPHM).

20. Cline Edmonds, interviewed by Mary E. Davidson, April 15, 1976 (PPHM); Mrs. Jimmie Cunningham, interviewed by Mary E. Davidson, February 19, 1976 (PPHM); C. W. Case, interviewed by Mary E. Davidson, November 19, 1975 (PPHM); Don Baker, interviewed by Mary E. Davidson, April 27, 1976 (PPHM); Mrs. Claude Ruby, interviewed by Mary E. Davidson, November 7, 1975 (PPHM); Horace Hickox, interviewed by Don Abbe, March 15, 1982 (PPHM); M. T. Johnson, interviewed by Marjorie L. Morris, June 27, 1972 (PPHM).

21. *Borger Daily Herald,* June 30, 1927; Ira M. Powell, "Great Church Comes out of Chaotic Conditions," *Texas Evangel* (October 1940): 12–15; Mrs. F. L. Cox and Mrs. Mary Moser, interviewed by Mary E. Davidson, May 20, 1976 (PPHM); Jo Ann Rumpel (reminiscence), n.d. (PPHM); John Edwards, interviewed by Bobby Weaver, June 26, 1982 (author's collection).

22. Mrs. Aline Flaugher, interviewed by Mary E. Davidson, April 13, 1976 (PPHM); George Finger, interviewed by Mary E. Davidson, November 19, 1975 (PPHM); Mrs. Jimmie Cunningham, interviewed by Mary E. Davidson, February 19, 1976 (PPHM); Frank Castleberry, interviewed by Mary E. Davidson, February 12, 1976 (PPHM); Don Baker, interviewed by Mary E. Davidson, April 27, 1976 (PPHM); Leonard Riley, interviewed by Mary E. Davidson, April 22, 1976 (PPHM); Fritz Thompson, interviewed by Mary E. Davidson, April 13, 1976 (PPHM); David Warren, interviewed by Mary E. Davidson, April 15, 1976 (PPHM); *Amarillo Daily News,* October 23, 1928.

23. *Amarillo Daily News,* October 8, 1927; October 20, 1927; October 23, 1927; July 19, 1928; *OW,* March 4, 1927, 39; June 14, 1929, 48; June 21, 1929, 49; June 28, 1929, 101; August 16, 1929, 54; October 25, 1929, 95.

24. N. D. Bartlett (reminiscence), July 31, 1936 (PPHM); M. K. Brown, interviewed by A. H. Doucette, August 21, 1944 (PPHM); Survey of *Annual Reports of Texas Railroad Commission, 1926–1940;* Survey of the weekly production reports in *OW,* January 7, 1927–December 30, 1935; U.S. Department of Commerce, Bureau of the Census, *Compendium of the Fourteenth Census of the United States, 1920, Vol. I, Number and Distribution of Inhabitants* (Washington, D.C.: G.P.O., 1921), 629–42; U.S. Department of Commerce, Bureau of the Census, *Compendium of the Fifteenth Census of the United States, 1930, Vol. I, Number and Distribution of Inhabitants* (Washington, D.C.: G.P.O., 1931), 1053–92.

25. C. W. Chase, interviewed by Mary E. Davidson, November 19, 1975 (PPHM); Ruth Clary Burrell, interviewed by Judy McKowan, November 11, 1975 (PPHM); C. Don Hughes, interviewed by Bobby Weaver, July 23, 1982 (PPHM); C. Don Hughes, interviewed by Amelia S. Nelson, November 4, 1979 (PPHM); F. Stanley, *The Early Days of the Oil Industry in the Texas Panhandle, 1919–1929* (Borger, Tex.: Hess, 1973), 282–85.

26. N. D. Bartlett (reminiscence), July 31, 1936 (PPHM); M. K. Brown, interviewed by A. H. Doucette, August 21, 1944 (PPHM); C. W. Case, interviewed by Mary E. Davidson November 19, 1975 (PPHM); Ruth Clay Burrell, interviewed by Judy McKowan, Novem-

ber 1, 1975 (PPHM); C. Don Hughes, interviewed by Bobby Weaver, July 23, 1982 (PPHM); C. Don Hughes, interviewed by Amelia S. Nelson, November 4, 1975 (PPHM).

27. Ernest O. Thompson, "Summary of the Development of the Panhandle Field under Conservation Regulations," *Panhandle-Plains Historical Review* 12 (1939): 13–23; *OW,* August 30, 1929, 54; September 6, 1929, 48; December 27, 1929, 51; February 7, 1930, 55; *Amarillo Daily News,* January 5, 1930.

28. Olive Elizabeth Henshaw Hills, interviewed by Cindy Hills Melanson, November 18, 1982 (PPHM); *Census Compendium, 1920,* 629–42; *Census Compendium, 1930,* 1053–92.

29. For a similar analysis of the boomtown phenomenon, see Gerald R. Forbes, "Southwestern Boom Towns," *Chronicles of Oklahoma* 17 (December 1939): 393–400, and Olien and Olien, *Oil Booms: Social Change in Five Texas Towns,* 1–18.

## CHAPTER 6

1. Albert Adams, interviewed by Bobby H. Johnson, August 12, 1970 (SFA); Olga Hermann Lapin, interviewed by Bobby H. Johnson, July 31, 1970 (SFA); Loyce Phillips, interviewed by Bobby H. Johnson, July 29, 1970 (SFA); Phillip A. Sanders, interviewed by Bobby H. Johnson, August 19, 1970 (SFA); Bill N. Taylor, interviewed by Bobby H. Johnson, August 13, 1970 (SFA).

2. *OW,* November 8, 1929, 28; Rister, *Oil: Titan of the Southwest,* 221–22; Nicholas George Malavis, *Bless the Pure and Humble* (College Station: Texas A&M University Press, 1996), 41–43.

3. Hinton and Olien, *Oil in Texas,* 169–70.

4. Roger M. Olien and Diana Davids Hinton, *Wildcatters: Texas Independent Oilmen* (College Station: Texas A&M University Press, 2007), 56.

5. E. C. Laster, interviewed by R. M. Hays, 1956 (CAH); Hinton and Olien, *Oil in Texas,* 169–70; Olien and Hinton, *Wildcatters,* 56.

6. E. C. Laster, interviewed by R. M. Hays, 1956 (CAH).

7. Albert Adams, interviewed by Bobby H. Johnson, August 12, 1970 (SFA); *Dallas Morning News,* October 15, 1930.

8. J. A. Rush, interviewed by Mody Boatright, September 11, 1956 (CAH); P. W. McFarland, "East Texas Oil Field," *Bulletin of the American Association of Petroleum Geologists* 15 (January–December 1931), part 2, 843.

9. Mrs. E. H. Spear, interviewed by Bobby H. Johnson, August 13, 1970 (SFA).

10. L. D. Winfrey, interviewed by Mrs. Robert A. Montgomery, August 2, 1959 (CAH).

11. A. C. Hopper, interviewed by Bobby H. Johnson, August 8, 1970 (SFA).

12. L. D. Winfrey, interviewed by Mrs. Robert A. Montgomery, August 2, 1959 (CAH); N. L. Field, interviewed by Bobby H. Johnson, August 20, 1970 (SFA); Mrs. N. L. Field, interviewed by Bobby H. Johnson, August 20, 1970 (SFA).

13. N. L. Field, interviewed by Bobby H. Johnson, August 20, 1970 (SFA).

14. Harold R. Johnson, interviewed by Bobby H. Johnson, August 15, 1970 (SFA); for an in-depth discussion of Humble's personnel policies concerning wages, hours, and benefits during the 1930s, see Henrietta M. Larson and Kenneth Wiggins Porter, *History of Humble Oil and Refining Company* (New York: Harpers and Brothers, 1959), 355–62.

15. Larson and Porter, *History of Humble Oil,* 384–85.

16. Homer J. "Jack" Davis, interviewed by Bobby H. Johnson, August 21, 1970 (SFA); Harold R. Johnson, interviewed by Bobby H. Johnson, August 15, 1970 (SFA); B. H. Smith, interviewed by Bobby H. Johnson, 1970 (SFA); Joe D. Lacy, interviewed by Bobby H. Johnson, August 15, 1970 (SFA); Charlotte Baker Montgomery, interviewed by Bobby H. Johnson, July 16, 1985 (SFA).

17. Capt. B. C. Baldwin, interviewed by Bobby H. Johnson, September 19, 1970 (SFA); Mrs. H. E. Sherwood, interviewed by Bobby H. Johnson, 1970 (SFA); Mrs. Sam W. Ross, interviewed by Bobby H. Johnson, 1970 (SFA); Mrs. L. L. Tullos Skeeters, interviewed by Bobby H. Johnson, 1970 (SFA); Albert Adams, interviewed by Bobby H. Johnson, August 12, 1970 (SFA); L. D. Winfrey, interviewed by Mrs. Robert A. Montgomery, August 2, 1959 (CAH).

18. Crown Dixon, interviewed by Bobby H Johnson, August 21, 1970 (SFA); Mary Flory Love, interviewed by Bobby H. Johnson, July 30, 1970 (SFA); Mrs. Eugene C. Kennedy, interviewed by Bobby H. Johnson, 1970 (SFA); Phillip A. Sanders, interviewed by Bobby H. Johnson, August 19, 1970 (SFA).

19. Mary Flory Love, interviewed by Bobby H. Johnson, July 30, 1970 (SFA); Mrs. A. C. Hopper, interviewed by Bobby H. Johnson, August 8, 1970 (SFA).

20. Charles A. Casey, interviewed by Bobby H. Johnson, August 7, 1970 (SFA); Mrs. L. L. Tullos Skeeters, interviewed by Bobby H. Johnson, 1970 (SFA); Lynch, *Roughnecks, Drillers, and Tool Pushers,* 55.

21. Mrs. L. L. Tullos Skeeters, interviewed by Bobby H. Johnson, 1970 (SFA).

22. Joe D. Lacy, interviewed by Bobby H. Johnson, August 15, 1970 (SFA); Mrs. Eugene Kennedy Elder, interviewed by Bobby H. Johnson, 1970 (SFA); A. C. Hopper, interviewed by Bobby H. Johnson, August 8, 1970 (SFA); Mrs. L. L. Tullos Skeeters, interviewed by Bobby H. Johnson, 1970 (SFA); Dr. James L. Nichols, interviewed by Bobby H. Johnson, August 11, 1970 (SFA); Mrs. A. C. Hopper, interviewed by Bobby H. Johnson, August 8, 1970 (SFA).

23. A. C. Hopper, interviewed by Bobby H. Johnson, August 8, 1970 (SFA); Charles A. Casey, interviewed by Bobby H. Johnson, August 7, 1970 (SFA); Bill N. Taylor, interviewed by Bobby H. Johnson, August 18, 1970 (FSA); Mrs. Gladys B. Foshee, interviewed by Bobby H. Johnson, August 30, 1970 (FSA); Captain B. C. Baldwin, interviewed by Bobby H. Johnson, September 9, 1970 (FSA); Albert Adams, interviewed by Bobby H. Johnson, August 13, 1970 (FSA).

24. Lynch, *Roughnecks, Drillers, and Tool Pushers,* 62; Mrs. L. L. Tullos Skeeters,

interviewed by Bobby H. Johnson, 1970, (SFA); Mrs. Olga Herrmann Lapin, interviewed by Bobby H. Johnson, July 31, 1970 (SFA); Charlotte Baker Montgomery, interviewed by Bobby H. Johnson, July 16, 1985 (SFA); Mrs. H. E. Sherwood, interviewed by Bobby H. Johnson, 1970 (SFA).

25. *OW,* February 13, 1931, 63; August 28, 1931, 14; September 11, 1931, 17; September 19, 1932, 43; February 6, 1933, 6; Rister, *Oil: Titan of the Southwest,* 315–26.

26. *OW,* June 20, 1932, 15; February 6, 1933, 28; Rister, *Oil: Titan of the Southwest,* 322.

27. Rister, *Oil: Titan of the Southwest,* 313–25; Hinton and Olien, *Oil in Texas,* 180–81; for an in-depth view of the entire series of events surrounding the East Texas oilfield legal situation during this period, consult Malavis, *Bless the Pure and Humble.*

28. L. D. Winfrey, interviewed by Mrs. Robert A. Montgomery, August 2, 1959 (CAH); Raymond A. Robinson, interviewed by Bobby H. Johnson, August 20, 1970 (SFA); Mrs. L. L. Tullos Skeeters, interviewed by Bobby H. Johnson, 1970 (SFA); Joe D. Lacy, interviewed by Bobby H. Johnson, August 5, 1970 (SFA); N. L. Field, interviewed by Bobby H. Johnson, August 20, 1970 (SFA); Loyce Phillips, interviewed by Bobby H. Johnson, July 29, 1970 (SFA); Brantly, *History of Oil Well Drilling,* 353, 360, 1150–53; Lynch, *Roughnecks, Drillers, and Tool Pushers,* 42; *OW,* April 9, 1926, 51; Wallace Davis, "Weight Indicator Solves Many Drilling Problems," *OW,* March 8, 1929, 40–46.

29. Lynch, *Roughnecks, Drillers, and Tool Pushers,* 15; Louis C. Sands, "Oil Field Development and Petroleum Production," in *Handbook of the Petroleum Industry,* ed. David T. Day (New York: John Wiley and Sons, 1921), 1:260–62; Brantly, *History of Oil Well Drilling,* 1060–84, provides a succinct overview of the technological history of the development of the rotary drill bit with particular emphasis on the Hughes rotary cone developments.

30. Stanley Gill, "A Study of Rotary Drilling Fluid," *OW,* June 13, 1932, 24–36. This report on the results of the API study gives a succinct analysis of the prejudices against rotary drilling in regard to the use of mud and then exposes the fallacies of those arguments.

31. Charlie Storms, interviewed by Ned Dewitt, August 16, 1939, in Lambert and Franks, eds., *Voices from the Oil Fields,* 22–38; L. D. Winfrey, interviewed by Mrs. Robert A. Montgomery, August 2, 1959 (CAH); *OW,* May 31, 1919, 30.

32. Lynch, *Roughnecks, Drillers, and Tool Pushers,* 21–22; Jimmy Zeigler, interviewed by Bobby Weaver, 1957 (author's collection); L. D. Winfrey, interviewed by Mrs. Robert A. Montgomery, August 2, 1959 (CAH).

33. Kirk Kite, s.v. "Highway Development," in *Handbook of Texas Online;* John D. Huddleston, s.v. "Texas Department of Transportation," in *Handbook of Texas Online;* Loyce Phillips, interviewed by Bobby H. Johnson, July 29, 1970 (SFA); N. L. Field, interviewed by Bobby H. Johnson, August 20, 1970 (SFA); Mrs. L. L. Tullos Skeeters, interviewed by Bobby H. Johnson, 1970 (SFA); Olga Hermann Lapin, interviewed by Bobby H. Johnson, July 31, 1970 (SFA); A. C. Hopper, interviewed by Bobby H. Johnson, August 8, 1970 (SFA).

## CHAPTER 7

1. Charles D. Vertrees, s.v. "The Permian Basin," in *Handbook of Texas Online.*

2. Frank T. Pickrell, interviewed by Berte Haigh, June 4, 1969 (PBPM).

3. Harold T. Morely, interviewed by Samuel D. Myres, October 20,1970 (PBPM).

4. Harold T. Morely, interviewed by Samuel D. Myres, October 20, 1970 (PBPM); Dee Locklin, interviewed by Samuel D. Myres, February 13, 1970 (PBPM); P. O. Sill, interviewed by Samuel D. Myres, May 12, 1970 (PBPM); George W. Ramer, interviewed by Samuel D. Myres, February 12, 1970 (PBPM).

5. George T. Abell, Charles Vertrees, and Berte Haigh, interviewed by Samuel D. Myres, February 2, 1971 (PBPM); T. B. "Buck" Harris, interviewed by Bobby Weaver, June 15, 1978 (SWC); W. W. Allman, interviewed by Samuel D. Myres, June 19, 1970 (PBPM); F. Arthur Stout, interviewed by Samuel D. Myres, 1970 (PBPM).

6. Frank T. Pickrell, interviewed by Berte Haigh, June 4, 1969 (PBPM); Dee Locklin, interviewed by Samuel D. Myres, February 13, 1970 (PBPM).

7. Frank T. Pickrell, interviewed by Berte Haigh, June 4, 1969 (PBPM); although this is the best-known and most romantic story of the naming of the Santa Rita #1 there are several other versions. Some of those stories were even given by Frank Pickrell in several interviews over the years. A plausible variant of the story can be found in Julia Cauble Smith, "The Early Development of the Big Lake Field, Reagan County, Texas" (master's thesis, University Texas of the Permian Basin, 1986), 16.

8. Dee Locklin, interviewed by Samuel D. Myres, February 13, 1970 (PBPM).

9. Clayton W. Williams Sr., interviewed by Samuel D. Myres, August 21, 1969 (PBPM); Dee Locklin, interviewed by Samuel D. Myres, February 13, 1970 (PBPM).

10. Dee Locklin, interviewed by Samuel D. Myres, February 13, 1970 (PBPM).

11. Frank T. Pickrell, interviewed by Berte Haigh, June 4, 1969 (PBPM); for a thorough overview of the discovery and subsequent development of the Big Lake oilfield, see Smith, "Early Development."

12. H. A. Hedberg, interviewed by Samuel D. Myres, June 26, 1970 (PBPM); W. H. "Bill" Collyns, interviewed by Richard Mason, January 25, 1980 (SWC).

13. W. H. "Bill" Collyns, interviewed by Richard Mason, January 25, 1980 (SWC); Elizabeth Lee, interviewed by Paula Marshall-Gray, 2004 (Paula Marshall-Gray personal collection); Paula Marshall-Gray, "Texon, Texas: Forging a Model Company Town in the Permian Basin," *Permian Historical Annual,* vol. 46, 2006; for the most comprehensive account of Texon and the social conditions of the town, see Paula Marshall-Gray, "Paradise Lost? Paradise Found? A Re-Collective History of the Oilfield Culture in a West Texas Company Town," (PhD diss., Texas Tech University, 2008).

14. William W. Allman, interviewed by Samuel D. Myres, June 19, 1970 (PBPM); T. B. "Buck" Harris, interviewed by Bobby Weaver, June 15, 1978 (SWC); Thomas C. Hogan, interviewed by Paul Patterson, May 1, 1968 (SWC); James "Pete" Williams, interviewed by Samuel D. Myres, September 15, 1971 (PBPM).

15. Cullen Akins, interviewed by Samuel D. Myres, June 17, 1970 (PBPM); Samuel D. Myres, *The Permian Basin: Petroleum Empire of the Southwest,* vol. 1, *Era of Discovery from the Beginning to the Depression* (El Paso: Permian Press, 1973), 65–69, 344–49; Rister, *Oil: Titan of the Southwest,* 292–300.

16. U.S. Department of Commerce, Bureau of the Census, *Census of Population, 1920.*

17. E. P. "Jack" Rainoseck, interviewed by Samuel D. Myres, June 17, 1970 (PBPM); F. N. Beane, interviewed by Samuel D. Myres, June 19, 1970 (PBPM); F. Arthur Stout, interviewed by Samuel D. Myres, 1970 (PBPM); F. E. "Ellis" Summers, interviewed by Richard Mason, March 31, 1981 (SWC); Clyde Barton, interviewed by Samuel D. Myres, July 15, 1970 (PBPM); Jim S. Peebles, interviewed by Samuel D. Myres, July 16, 1970 (PBPM); Lela Laughlin, interviewed by Richard Mason, May 20, 1981 (SWC); Joseph W. Graybeal, interviewed by Samuel D. Myres, 1970 (PBPM).

18. W. W. Allman, interviewed by Samuel D. Myres, June 19, 1970 (PBPM); F. E. "Ellis" Summers, interviewed by Richard Mason, March 31, 1981 (SWC); Estha Briscoe Stowe, *Oil Field Child* (Fort Worth: Texas Christian University Press Press, 1989), 65–69; for a more detailed overview of oil company camps during this period, see Diana Davids Hinton, "Creating Company Culture: Oil Company Camps in the Southwest, 1920–1960," *Southwestern Historical Quarterly* 111, no. 4 (2008): 368–87.

19. E. P. "Jack" Rainoseck, interviewed by Samuel D. Myres, June 17, 1970 (PBPM); E. N. Beane, interviewed by Samuel D. Myres, June 19, 1970 (PBPM); Hinton, "Creating Company Culture," 379; William Wolf, interviewed by Samuel D. Myres, February 12, 1970 (PBPM); J. Ben Carsey, interviewed by Samuel D. Myres, October 9, 1970 (PBPM).

20. Pete Sitton, interviewed by Richard Mason, September 21, 1981 (SWC); W. R. Johnson, interviewed by J. M. Skaggs, April 4, 1968 (SWC); Clayton W. Williams, interviewed by Samuel D. Myres, August 21, 1969 (PBPM); George W. Ramer, interviewed by Samuel D. Myres, February 12, 1970 (PBPM).

21. F. Arthur Stout, interviewed by Samuel D. Myres, n.d. (PBPM); Thomas C. Hogan, interviewed by Paul Patterson, May 1, 1968 (SWC); S. O. Cooper, interviewed by Samuel D. Myres, May 13, 1970 (PBPM).

22. T. B. "Buck" Harris, interviewed by Bobby Weaver, June 15, 1978 (SWC); Cullen Akins, interviewed by Samuel D. Myres, June 17, 1970 (PBPM); William Wolf, interviewed by Samuel D. Myres, February 12, 1970 (PBPM); Clyde Barton, interviewed by Samuel D. Myres, July 15, 1970 (PBPM); W. W. Allman, interviewed by Samuel D. Myres, June 19, 1970 (PBPM); *OW,* February 17, 1928, 53; February 27, 1928, 24.

23. U.S. Department of Commerce, Bureau of the Census, *Census of Population, 1920,* indicates that there were only eighty-one people in Winkler County in 1920; Clyde Barton, interviewed by Samuel D. Myres, July 16, 1970 (PBPM); *Odessa Times,* November 4, 1927; *Pecos Enterprise and Gusher,* July 23, 1926; Clarence Pope, *An Oil Scout in the Permian Basin* (El Paso: Permian Press, 1972), 63; Olien and Olien, *Oil Booms: Social Change in Five Texas Towns,* 15–16.

24. J. C. Avary, interviewed by Richard Mason, April 1, 1981 (SWC); Jim S. Peeples, interviewed by Samuel D. Myres, July 16, 1970 (PBPM); Pete Sitton, interviewed by Richard Mason, September 21, 1981 (SWC); W. R. Johnson, interviewed by J. M. Skaggs, April 14, 1968 (SWC); T. B. "Buck" Harris, interviewed by Bobby Weaver, June 15, 1978 (SWC); Carl Weaver, interviewed by Paul Patterson, July 5, 1966 (SWC); Tony Wilburn, interviewed by Richard Mason, April 1, 1981 (SWC).

25. C. L. McKinney, interviewed by Samuel D. Myres, September 21, 1971 (PBPM); George Donnely, interviewed by Samuel D. Myres, February 1971 (PBPM); E. E. Brackens, interviewed by Samuel D. Myres, July 16, 1970 (PBPM).

26. Joseph W. Graybeal, interviewed by Samuel D. Myres, n.d. (PBPM); Jim S. Peebles, interviewed by Samuel D. Myres, July 16, 1970 (PBPM); Pete Sitton, interviewed by Richard Mason, September 21, 1981 (SWC); W. R. Johnson, interviewed by J. M. Skaggs, April 14, 1968 (SWC); C. L. McKinney, interviewed by Samuel D. Myres, September 21, 1971 (PBPM).

27. Olien and Olien, *Oil Booms: Social Change in Five Texas Towns*, 133–40, provides an excellent description of the prevalence of criminal activity in Wink from the time of its founding into the 1930s.

28. F. G. "Ellis" Summers, interviewed by Richard Mason, March 31, 1981 (SWC).

29. In Frank Mangan, *The Pipeliners* (El Paso: Guynes Press, 1977), 57.

30. *OW*, July 30, 1926, 88; George W. Ramer, interviewed by Samuel D. Myres, February 12, 1970 (PBPM); Clyde Barton, interviewed by Samuel D. Myres, July 15, 1970 (PBPM).

31. "Review of Petroleum Development in West Texas," *Petroleum Engineer* (July 1, 1936): 133–43; J. Ben Carsey, interviewed by Samuel D. Myres, October 9, 1970 (PBPM); W. H. "Bill" Collyns, interviewed by Richard Mason, January 25, 1980 (SWC).

32. Olien and Olien, *Oil Booms: Social Change in Five Texas Towns*, 176–77; United States Department of Commerce, Bureau of the Census, *Census of Population, 1920, 1930, and 1940.*

33. Samuel D. Myres, *The Permian Basin: Petroleum Empire of the Southwest*, vol. 2, *Era of Advancement, from the Depression to the Present* (El Paso: Permian Press, 1977), 84–86; T. B. "Buck" Harris, interviewed by Bobby Weaver, June 15, 1978 (SWC); W. C. "Bill" Hunt, interviewed by Bobby Weaver, December 2, 1977 (SWC); Pete Sitton, interviewed by Richard Mason, September 21, 1981 (SWC); Jim S. Peebles, interviewed by Samuel D. Myres, July 16, 1970 (PBPM); Albert Whatley, interviewed by Samuel D. Myres, July 18, 1970 (PBPM); George R. Bentley, interviewed by Samuel D. Myres, May 20, 1970 (PBPM); William Wolf, interviewed by Samuel D. Myres, July 12, 1970 (PBPM); E. N. Beane, interviewed by Samuel D. Myres, June 19, 1970 (PBPM).

34. Cullen Akins, interviewed by Samuel D. Myres, June 17, 1970 (PBPM); George Ramey, interviewed by Samuel D. Myres, February 12, 1970 (PBPM); Jim S. Peebles, interviewed by Samuel D. Myres, July 16, 1970 (PBPM); Brantly, *History of Oil Well Drilling*, 279–84, 360–63, 1083–87.

35. Burk Paschall, interviewed by Mody Boatright, July 30, 1952 (CAH); Pete Sitton, interviewed by Richard Mason, September 21, 1981 (SWC); Bruce Ashburn, interviewed by Paul Patterson, April 29, 1968 (SWC); H. C. "Doc" Cotton, interviewed by Bobby Weaver, November 7, 1978 (SWC); Thomas C. Hogan, interviewed by Paul Patterson, May 11, 1968 (SWC).

36. Larson and Porter, *History of Humble Oil,* 352.

37. Clarence "Sonny" Keith, interviewed by Bobby Weaver, 1957 (author's collection).

38. Larson and Porter, *History of Humble Oil,* 351–53; Bruce Ashburn, interviewed by Paul Patterson, April 29, 1968 (SWC); Brantly, *History of Oil Well Drilling,* 299: F. Arthur Stout, interviewed by Samuel D. Myres, n.d. (PBPM).

39. James C. Moroney, "Oilfield Strike of 1917," in *Handbook of Texas,* ed. Ron Tyler et al., 4:1119; *OW,* November 24; December 1; December 29, 1917; February 2, 1918; William Wolf, interviewed by Samuel D. Myres, February 12, 1970 (PBPM); *OW,* January 28, 1927, 46; February 11, 1927, 31; July 6, 1936; Pete Sitton, interviewed by Richard Mason, September 21, 1981 (SWC).

40. *OW,* November 16, 1936, 63; Lynch, *Roughnecks, Drillers, and Tool Pushers,* 128, 159–60.

41. H. E. "Eddie" Chiles, interviewed by Samuel D. Myres, February, 1971 (PBPM); C. L. McKinney, interviewed by Samuel D. Myres, September 23, 1971 (PBPM).

42. T. B. "Buck" Harris, interviewed by Bobby Weaver, June 15, 1978 (SWC); Cullen Akins, interviewed by Samuel D. Myres, June 17, 1970 (PBPM); Carl Weaver, interviewed by Paul Patterson, July 15, 1961 (SWC).

43. Cecil Bickley, interviewed by Richard Mason, August 24, 1982 (SWC); Rayford Fowler, interviewed by Richard Mason, February 2, 1982 (SWC); Buff and LaCosta Ivey, interviewed by Bobby Weaver, September 7, 1978 (SWC); Jack Akin, interviewed by Bobby Weaver, November 8, 1978 (SWC); Cecil Bickley, interviewed by Bobby Weaver, November 8, 1978 (SWC); H. C. "Doc" Cotton, interviewed by Bobby Weaver, November 7, 1978 (SWC).

44. Elliott Cowden, interviewed by Samuel D. Myres, September 16, 1970 (PBPM); George R. Bentley, interviewed by Samuel D. Myres, May 20, 1970 (PBPM); Clyde Barton, interviewed by Samuel D. Myres, July 15, 1970 (PBPM).

## CHAPTER 8

1. Hinton and Olien, *Oil in Texas,* 228–30.

2. *OW,* August 3, 1942, 30–32; February 1, 1943, 83: Larson and Porter, *History of Humble Oil,* 579.

3. A. R. McTee, "Manpower Looms as Oil's Most Critical Problem," *OW,* August 2, 1943, 20–22; Reaves Rivers, "Oil Industry Feeling Manpower Pinch," *OW,* October 4, 1942, 34–36; L. J. Logan, "Oil Industry and Oil Workers Highly Essential to the War,"

*OW*, July 5, 1943, 11–13; A. R. McTee, "Manpower Shortage Imperils Industry's Exploratory Work," *OW*, August 2, 1943, 20–22; *OW*, September 14, 1942, 15–16; January 24, 1944, 8–9; January 17, 1944, 9; October 4, 1942, 34–36; August 9, 1943, 30–32; Hinton and Olien, *Oil in Texas*, 225.

4. Hinton and Olien, *Oil in Texas*, 231.

5. J. D. Brown, interviewed by Bobby Weaver, 1960 (author's collection).

6. Ralph Thompson, interviewed by Bobby Weaver, 1965 (author's collection).

7. Buff Ivey, interviewed by Bobby Weaver, November 7, 1978 (SWC).

8. H. C. "Doc" Cotton, interviewed by Bobby Weaver, November 7, 1978 (SWC).

9. Pete Sitton, interviewed by Richard Mason, September 21, 1981 (SWC).

10. Larson and Porter, *History of Humble Oil*, 586–87; Rister *Oil: Titan of the Southwest*, 355–62.

11. "Whitey" Harding, interviewed by Bobby Weaver, 1958 (author's collection); Bill Walker, interviewed by Bobby Weaver, 1963 (author's collection); Rister, *Oil: Titan of the Southwest*, 359.

12. W. R. Johnson, interviewed by J. M. Skaggs, April 14, 1968 (SWC); Carl Weaver, interviewed by Paul Patterson, June 5, 1966 (SWC); *Odessa American*, September 2, 1942.

13. Olien and Olien, *Oil Booms: Social Change in Five Texas Towns*, 177; U.S. Department of Commerce, Bureau of the Census, *Census of Population, 1940, 1950.*

14. Carl Weaver, interviewed by Paul Patterson, June 5, 1966 (SWC).

15. Carl Weaver, interviewed by Paul Patterson, July 5, 1966 (SWC); James Roberts, interviewed by Jeff Townsend, August 22, 1972 (SWC); Floyd Peacock, interviewed by Jeff Townsend, August 22, 1972 (SWC).

16. *OW*, November 6, 1936, 63; October 26, 1942, 32; Lynch, *Roughnecks, Drillers, and Tool Pushers*, 128.

17. Brantly, *History of Oil Well Drilling*, 300–326.

18. Richard Whatley, interviewed by Bobby Weaver, 1965 (author's collection); P. O. Sill, interviewed by Samuel D. Myres, May 12, 1970 (PBPM); S. O. Cooper, interviewed by Samuel D. Myres, May 13, 1970 (PBPM); Pete Sitton, interviewed by Richard Mason, September 21, 1981 (SWC); Brantly, *History of Oil Well Drilling*, 300–326.

19. H. C. "Doc" Cotton, interviewed by Bobby Weaver, November 7, 1978 (SWC); Buff Ivey, interviewed by Bobby Weaver, November 7, 1978 (SWC); Robert A. "Bob" Cullen, interviewed by Bobby Weaver, July 30, 2009 (author's collection); Claude E. "Peeper" Johnson, interviewed by Bobby Weaver, July 29, 2009 (author's collection).

20. Pete Sitton, interviewed by Richard Mason, September 21, 1981 (SWC); Robert A. "Bob" Cullen, interviewed by Bobby Weaver, July 30, 2009 (author's collection); Claude E. "Peeper" Johnson, interviewed by Bobby Weaver, July 29, 2009 (author's collection).

21. Robert A. "Bob" Cullen, interviewed by Bobby Weaver, July 30, 2009 (author's collection)

22. Robert A. "Bob" Cullen, interviewed by Bobby Weaver, July 30, 2009 (author's

collection); Jerry Holt, interviewed by Bobby Weaver, June 10, 1980 (author's collection); Jerry Duke, interviewed by Bobby Weaver, October 14, 1992 (author's collection); Claude "Peeper" Johnson, interviewed by Bobby Weaver, July 29, 2009 (author's collection).

23. Samuel D. Myres, *The Permian Basin: Era of Advancement* (El Paso: Permian Press, 1977), 33, 278–81.

24. *Snyder Daily News,* December 9, 1962, October 7, 1973; Buff and LaCosta Ivey, interviewed by Bobby Weaver, November 7, 1978 (SWC); K. O. Pitner, interviewed by Jeff Townsend, August 22, 1972 (SWC); L. E. Griffen, interviewed by Jeff Townsend, August 22, 1972 (SWC); Olien and Olien, *Oil Booms: Social Change in Five Texas Towns,* 17, 177.

25. Myres, *Permian Basin: Era of Advancement,* 289.

26. Myres, *Permian Basin: Era of Advancement,* 289; Matt Ware, interviewed by Bobby Weaver, July 23, 2009 (author's collection); Bob Cullen, interviewed by Bobby Weaver, July 30, 2009 (author's collection).

27. Herbert "Peewee" Johnson, interviewed by Bobby Weaver, 1959 (author's collection); Herman "Little Red" Michaels, interviewed by Bobby Weaver, 1958 (author's collection); Robert S. "Bob" Weaver, interviewed by Bobby Weaver, 1974 (author's collection); Matt Ware, interviewed by Bobby Weaver, July 23, 2009 (author's collection).

28. Myres, *Permian Basin: Era of Advancement,* 33–35, 373–78.

29. Matt Ware, interviewed by Bobby Weaver, July 23, 2009 (author's collection); Robert S. "Bob" Weaver, interviewed by Bobby Weaver, 1974 (author's collection); J. P. Judkins, interviewed by Bobby Weaver, 1961 (author's collection)

30. Matt Ware, interviewed by Bobby Weaver, July 23, 2009 (author's collection).

31. Olien and Olien, *Oil Booms: Social Change in Five Texas Towns,* 14, 177–78; U.S. Department of Commerce, Bureau of the Census, *Census of Population: 1950, 1960.*

32. Robert S. "Bob" Weaver, interviewed by Bobby Weaver, 1974 (author's collection); Lynch, *Roughnecks, Drillers, and Tool Pushers,* 128, 159–60; Survey of *Lloyd's City Directories: Odessa, Texas,* for the years 1946–60.

33. James Roberts, interviewed by Jeff Townsend, August 22, 1972 (SWC); Floyd Peacock, interviewed by Jeff Townsend, August 22, 1972 (SWC); W. C. "Bill" Hunt, interviewed by Bobby Weaver, December 12, 1977 (SWC); E. P. "Jack" Rainosek, interviewed by Samuel D. Myres, June 17, 1970 (PBPM); Hinton, "Creating Company Culture," 386.

34. Jimmy Zeigler, interviewed by Bobby Weaver, 1958 (author's collection).

35. Herbert L. "Peewee" Johnson, interviewed by Bobby Weaver, 1959 (author's collection).

36. In the year 2000 I did an informal survey of all company's connected in any way with tank building in the Midland/Odessa area, and all I could locate were some welding plants going under the name of tank companies and a variety of roustabout companies that did patchwork on tanks. I was assured by several of them that the last operating tank crews they could remember dated from the mid-1970s.

37. C. L. McKinney, interviewed by Samuel D. Myres, September 23, 1971 (PBPM).

## CHAPTER 9

1. Olien and Olien, *Oil Booms: Social Change in Five Texas Towns,* 1–23; Rister, *Oil: Titan of the Southwest,* 48–65.

2. Jack Knight, interviewed by Mody Boatright, September 1952 (CAH); Bruce Ashburn, interviewed by Paul Patterson, April 29, 1968 (SWC); Jim S. Peebles, interviewed by Samuel D. Myres, August 16, 1970 (PBPM); William Franklin "Hot Shot" Ash, interviewed by Roger Olien, March 2, 1979 (Roger Olien collection).

3. Alexander Balfour Patterson, interviewed by William A. Owens, August 4, 1953 (CAH); Sam Webb, interviewed by William A. Owens, September 1952 (CAH); Benjamin "Bud" Coyle, interviewed by William A. Owens, August 28, 1953 (CAH).

4. Larson and Porter, *History of Humble Oil,* 66–71, 198–208, 211–13; James C. Maroney, s.v. "Oilfield Strike of 1917," in *Handbook of Texas Online.*

5. *OW,* February 18, 1919, 32.

6. *OW,* October 5, 1918, 8.

7. *OW,* October 25, 1919, 43.

8. *OW,* October 5, 1918, 36; October 23, 1920, 171; January 21, 1922, 101; August 18, 1923, 9; January 19; 1924, 38; October 2, 1925, 188; July 24, 1920, 11; October 23, 1920, 195; November 23, 1918, 7; February 22, 1929, 111.

9. R. L. Dudley, "The Oil Worker Cared For," *OW,* October 23, 1920, 115–20.

10. *OW,* October 23, 1920, 115; M&M Thomas C. Hogan, interviewed by Paul Patterson, May 1, 1968 (SWC); William W. Allman, interviewed by Samuel D. Myres, June 19, 1970 (PBPM); E. P. "Jack" Rainoseck, interviewed by Samuel D. Myres, June 19, 1970 (PBPM); Homer J. "Jack" Davis, interviewed by Bobby H. Johnson, August 21, 1970 (SFA); M&M D. W. Page, interviewed by Mary E. Davidson, April 22, 1976 (PPHM); George Finger, interviewed by Mary E. Davidson, November 19, 1975 (PPHM) .

11. Alexander Balfour Patterson, interviewed by William A. Owens, August 4, 1953 (CAH).

12. A. R. McTee, "How Big Oil Companies Care for Employees," *OW,* February 9, 1924, 25–27, 47–48; J. F. Carter Jr., "Oil Company Builds Model Town," *OW,* April 26, 1924, 39–42; W. V. Gross, "Camp Beaty Cures That Tired Feeling," *OW,* November 14, 1924, 49.

13. Brantly, *History of Oil Well Drilling,* 15.

14. F. Arthur Stout, interviewed by Samuel D. Myres, n.d. (PBPM); Dan Lively, interviewed by Paul Patterson, April 23, 1968 (SWC).

15. Brantly, *History of Oil Well Drilling,* 15, 353–54, 359.

16. Curt G. Hamill, interviewed by William A. Owens, August 17, 1952 (CAH); Brantly, *History of Oil Well Drilling,* 326–27.

17. Lynch, *Roughnecks, Drillers, and Tool Pushers,* 22–23; J. A. Rush, interviewed by Mody Boatright, September 11, 1956 (CAH); James Donohoe, interviewed by William A.

Owens, August 1, 1952 (CAH); Jimmy Zeigler, interviewed by Bobby Weaver, 1958 (author's collection).

18. S. O. Cooper, interviewed by Samuel D. Myres, May 13, 1970 (PBPM); W. R. Johnson, interviewed by J. M. Skaggs, April 14, 1968 (SWC); P. O. Sill, interviewed by Samuel D. Myres, May 12, 1970 (PBPM).

19. C. D. Walters, "Advantages of Standardized Rigs, Derricks, and Rig Irons," *OW,* December 9, 1927, 23–24, 71; Brantly, *History of Oil Well Drilling,* 372.

20. Wallace Davis, "Weight Indicator Solves Many Drilling Problems," *OW,* March 8, 1929, 23; Wendell M. Jones, "Equipment for Drilling Exploratory Wells by the Rotary Method," *OW,* January 11, 1929, 86; Brantly, *History of Oil Well Drilling,* 1068, 1084.

21. *OW,* August 15, 1932, 17; Brantly, *History of Oil Well Drilling,* 19–20, 327; Gus Carey, interviewed by Fred Carpenter, August 9, 1972 (SWC).

22. Brantly, *History of Oil Well Drilling,* 327; Alexander Balfour Patterson, interviewed by William A. Owens, August 4, 1953 (CAH); Carl Angstadt, interviewed by Mody Boatright, August 4, 1952 (CAH).

23. *OW,* January 28, 1927, 46; February 11, 1927, 31; July 6, 1936, 29; Lambert and Franks, eds., *Voices from the Oil Fields,* 22–38; Brantly, *History of Oil Well Drilling,* 343–44.

24. Richard Whatley, interviewed by Bobby Weaver, 1965 (author's collection); Brantly, *History of Oil Well Drilling,* 1356–57; E. E. Brackens, interviewed by Samuel D. Myres, July 16, 1970 (PBPM).

25. Rister, *Oil: Titan of the Southwest,* 47; Claude L. Witherspoon, interviewed by Mody Boatright, June 27, 1953 (CAH).

26. Rister, *Oil: Titan of the Southwest,* 62–63.

27. Curt Hamill, interviewed by William A. Owens, July 17, 1953 (CAH).

28. Frank Dunn, interviewed by William A. Owens, February 23, 1955 (CAH).

29. *Gulf Coast Oil News,* January 16, 1917, 1.

30. *Gulf Coast Oil News,* March 17, 1917, 7; *OW,* March 11, 1919, 32; December 2, 1922, 46.

31. *Gulf Coast Oil News,* July 21, 1917, 17–21.

32. C. P. Bowie, Bulletin #55 of the Department of the Interior, Bureau of Mines, "Oil Storage Tanks and Reservoirs, with a Brief Discussion of Losses of Oil in Storage and Methods of Prevention," as cited in *OW,* April 20, 1918, 24.

33. *OW,* August 16, 1919, 32.

34. *OW,* December 15, 1917, 15; June 22, 1918, 26.

35. *OW,* July 28, 1917, 16–17; August 25, 1917, 32.

36. *OW,* July 28, 1917, 32; January 26, 1918, 26; June 14, 1919, 32.

37. *OW,* January 27, 1928, 24; William Wolf, interviewed by Samuel D. Myres, February 12, 1970 (PBPM); Clyde Barton, interviewed by Samuel D. Myres, July 15, 1970 (PBPM); H. L. Kauffman, "Some Facts Worth Knowing about Oil Storage," *OW,* March 15, 1929, 19–22.

38. Brad Mills, "Welded Tanks Becoming Popular in California," *OW,* March 14, 1930, 211; *OW,* February 26, 1921, 33; August 12, 1922, 11; March 1, 1924, 5; March 8, 1929, 81; October 11, 1929, 166.

39. *Gulf Coast Oil News,* April 14, 1917, 7; *OW,* May 22, 1920, 5; October 23, 1920, 14; May 2, 1930, 81; June 15, 1936, 20: Penn Ware, interviewed by Bobby Weaver, 1955 (author's collection); Bill Walker, interviewed by Bobby Weaver, 1963 (author's collection).

40. *Gulf Coast Oil News,* September 22, 1917, 9; February 23, 1918, 33.

41. *Gulf Coast Oil News,* September 22, 1917, 12; December 22, 1917, 12; *OW,* August 24, 1918, 13; February 26, 1921, 33; April 29, 1922, 1; August 12, 1922, 11; October 7, 1922, 1.

42. "Whitey" Harding, interviewed by Bobby Weaver, 1958 (author's collection); "Junior" Morton, interviewed by Bobby Weaver, 1958 (author's collection); Herbert Lee "Peewee" Johnson, interviewed by Bobby Weaver, 1959 (author's collection).

43. Penn Ware, interviewed by Bobby Weaver, 1955 (author's collection); Matt Ware, interviewed by Bobby Weaver, July 24, 2009 (author's collection); Robert Samuel "Bob" Weaver, interviewed by Bobby Weaver, 1974, 1984, 1990 (author's collection).

44. H. H. King, "Gulf Company Builds Long Line to Serve West Texas," *OW,* September 23, 1927, 76; *OW,* October 18, 1919, 208; Burt Hull, interviewed by William A. Owens, August 24, 1953 (CAH).

45. *OW,* July 14, 1923, 74; June 24, 1924, 31.

46. H. L. Kauffman, "Developments in Pipeline Welding Practices and Equipment," *OW,* March 14, 1930, 248–51; *OW,* January 31, 1930, 28; Alfred M. Leeston, "An Industry Is Born," in Alfred M. Leeston, John A. Crichton, and John C. Jacobs, *The Dynamic Natural Gas Industry* (Norman: University of Oklahoma Press, 1963), 8–10.

47. Ball, *This Fascinating Oil Business,* 179–82; John C. Jacobs, "From Field to Market," in Leeston, Crichton, and Jacobs, *Dynamic Natural Gas Industry,* 90–92; A. R. McTee, "Better Methods Hasten Pipe Line Building," *OW,* September 23, 1927, 95–100; *OW,* January 9, 1931, 53, 77–78.

48. Michael Leslie, interviewed by Bobby Weaver, September 4, 2008 (author's collection).

49. Kirk Kite, s.v. "Highway Development," in *Handbook of Texas Online;* John D. Huddleston, s.v. "Texas Department of Transportation," in *Handbook of Texas Online;* John D. Huddleston, "Highway Development: A Concrete History of Twentieth Century Texas," in *Texas: A Sesquicentennial Celebration* (Austin: Eakin Press, 1984), 254–56.

50. *OW,* June 1, 1918, 20.

51. *OW,* October 18, 1919, 83–86.

52. *OW,* November 12, 1918, 1; November 23, 1918, 40; June 21, 1919, 20; September 20, 1919, 37.

53. *OW,* November 30, 1918, 42; November 2, 1918, 1.

54. *OW,* June 25, 1926, 42.

55. Plummer M. Barfield, interviewed by William A. Owens, August 1, 1952 (CAH); William Edward Cotton, interviewed by William A. Owens, May 23, 1956 (CAH); L. D.

Winfrey, interviewed by Mrs. Robert A. Montgomery, August 12, 1959 (CAH); Charlie Davis, interviewed by Richard Mason, January 15, 1981 (SWC); N. L. Field, interviewed by Bobby H. Johnson, August 20, 1970 (SFA).

56. Gow, *Roughnecks, Rock Bits and Rigs,* 125–28; Ball, *This Fascinating Oil Business,* 121–22.

57. F. Stanley, *The Tex Thornton Story* (Nazareth, Tex.: F. Stanley, 1975), 5; W. H. "Toby" Mendenhall, interviewed by Bobby Weaver, March 13, 1985 (PPHM); C. L. McKinney, interviewed by Samuel D. Myres, September 23, 1971 (PBPM).

58. Gow, *Roughnecks, Rock Bits and Rigs,* 125–28; C. L. McKinney, interviewed by Samuel D. Myres, September 23, 1971 (PBPM); W. H. "Toby" Mendenhall, interviewed by Bobby Weaver, March 13, 1985 (PPHM).

59. C. L. McKinney, interviewed by Samuel D. Myres, September 23, 1971 (PBPM); W. H. "Toby" Mendenhall, interviewed by Bobby Weaver, March 13, 1985 (PPHM); Johnny Waddington, interviewed by Richard Mason, January 4, 1980 (SWC).

60. W. H. "Toby" Mendenhall, interviewed by Bobby Weaver, March 13, 1985 (PPHM); *OW,* May 6, 1927, 37.

61. C. L. McKinney, interviewed by Samuel D. Myres, September 23, 1971 (PBPM).

62. C. L. McKinney, interviewed by Samuel D. Myres, September 23, 1971 (PBPM); W. H. "Toby" Mendenhall, interviewed by Bobby Weaver, March 13, 1985 (PPHM).

63. Boatright, *Folklore of the Oil Industry,* 107–11; Walter Cline, interviewed by Mody Boatright, August 8, 1952 (CAH); Ed. P. Matteson, interviewed by Mody Boatright, June 19, 1953 (CAH).

64. Frank Hamilton, interviewed by William A. Owens, August 29, 1952 (CAH).

65. A. H. Clough, "Dynamite Ends Gas Fire," *Petroleum Age* (1920): 32; O. G. Lawson, interviewed by Mody Boatright, July 28, 1952 (CAH); Stanley Frank, "He Fights the Biggest Fires," *Saturday Evening Post,* May 2, 1959, 36.

66. Stanley, *Tex Thornton Story,* 9–10.

67. *Amarillo News,* January 30, 1928; Boatright, *Folklore of the Oil Industry,* 113–14; N. D. Bartlett, "He Handles Sudden Death without Fear," *Amarillo News,* 1933, clippings file PPHM; Stanley, *Tex Thornton Story,* 4–12; *Amarillo Daily News,* April 7, 1927; April 10, 1927; April 15, 1927; June 7, 1927; June 10, 1927; October 1, 1927; January 30, 1928; August 13, 1928.

68. L. C. E. Bignell, "Rumanian Gas Fire Finally Conquered," *Oil and Gas Journal* (March 1932): 16; Stanley Frank, "He Fights the Biggest Fires," *Saturday Evening Post,* May 2, 1959, 36; for an anecdotal overview of Kinley's career, see Jesse D. Kinley, *Call Kinley: Adventures of an Oil Well Firefighter* (Chickasha, Okla.: n.p., 1966).

69. For an overview of the career of "Red" Adair, see Philip Singerman, *An American Hero: The Red Adair Story* (Boston: Little, Brown, 1990).

70. James William Kinnear, interviewed by William A. Owens, August 24, 1953 (CAH); Plummer M. Barfield, interviewed by William A. Owens, August 1, 1952 (CAH);

William Edward Cotton, interviewed by William A. Owens, May 23, 1956 (CAH); A. J. Thaman and J. H. Anderson, interviewed by Mrs. Maude Ross, January 29, 1960 (CAH).

71. O. W. Killiam, interviewed by William A. Owens, May 7, 1956 (CAH); A. J. Thaman and J. H. Anderson, interviewed by Mrs. Maude Ross, January 29, 1960 (CAH); Harold Halsey, interviewed by Mrs. Maude Ross, January 27, 1960 (CAH); Emilio Zamora, "The Failed Promise of Wartime Opportunity for Mexicans in the Texas Oil Industry," *Southwestern Historical Quarterly* 95 (January 1992).

## CHAPTER 10

1. The words and phrases presented in this essay are not meant to be an exhaustive study of the language of the oilfield. They represent terminology that I picked up from all those oilfield hands I grew up around, worked with, or met over a period of some sixty years. These words and phrases are designed to help the reader better understand some of the specialized terminology associated with oilfield labor in the hope that those terms might give some insight into the unique culture of oilfield trash. Lord, how I miss the patch!

# Bibliography


## PRIVATE PAPERS

Charles N. Gould Collection, Western History Collection, University of Oklahoma.

Carl Coke Rister Collection, Southwest Collection, Texas Tech University.

## PUBLIC DOCUMENTS

*Annual Report of the Texas Railroad Commission, 1920, 1930.*

*Bulletin of the American Association of Petroleum Geologists.*

*Report of the Adjutant General of the State of Texas for the Fiscal Year, Sept., 1929–Aug. 31, 1930.*

U.S. Department of Commerce, Bureau of the Census. *Census of Population, 1910, 1920, 1930, 1940.*

## BOOKS

Andrews County Heritage Committee. *Andrews County History, 1876–1978.* Andrews, Tex.: Andrews County Heritage Committee, 1978.

Ball, Max W. *This Fascinating Oil Business.* New York: Bobbs-Merrill, 1940.

Boatright, Mody C. *Folklore of the Oil Industry.* Dallas: Southern Methodist University Press, 1963.

Boatright, Mody C., and William A. Owens. *Tales from the Derrick Floor: A People's History of the Oil Industry.* Lincoln: University of Nebraska Press, 1970.

Brantly, J. E. *History of Oil Well Drilling.* Houston: Gulf Publishing Company, 1971.

Clark, James A., and Michel Halbouty. *The Last Boom.* New York: Random House, 1972.

———. *Spindletop.* New York: Random House, 1952.

Day, David T., ed. *Handbook of the Petroleum Industry.* New York: John Wiley and Sons, 1923.

*Derrick's Handbook of the Petroleum Industry.* Oil City, Pa.: Derrick Publishing Company, 1900.

Gallaway, W. E., E. Ewing, C. M. Garrett, N. Tyler, and G. G. Bebout. *Atlas of Major Texas Oil Reservoirs.* Austin: University of Texas Bureau of Economic Geology, 1983.

Gould, Charles N. *Covered Wagon Geologist.* Norman: University of Oklahoma Press, 1959.

Gow, Sandy. *Roughnecks, Rock Bits and Rigs: The Evolution of Oil Well Drilling Technology in Alberta, 1883–1970.* Calgary, Alberta: University of Calgary Press, 2005.

Greever, William S. *The Bonanza West: The Story of Western Mining Rushes, 1848–1900.* Norman: University of Oklahoma Press, 1963.

Hamill, Curtis G. *We Drilled Spindletop.* Houston: privately published, 1957.

Hinton, Diana Davids, and Roger M. Olien. *Oil in Texas.* Austin: University of Texas Press, 2002.

House, Boyce. *Oil Boom.* Caldwell, Idaho: Caxton Printers, 1941.

———. *Oil Field Fury.* San Antonio, Tex.: Naylor, 1954.

———. *Roaring Ranger.* San Antonio, Tex.: Naylor, 1951.

———. *Were You in Ranger?* Dallas: Tardy Publishing, 1935.

Jenkins, John H., and H. Gordon Frost. *I'm Frank Hamer: The Life of a Texas Peace Officer.* Austin: State House Press, 1993.

Jones, John P. "Slim." *Borger: The Little Oklahoma.* Meridian, Tex.: privately published, 1927.

Kansas City Star. *World's Work.* Kansas City: Star, 1927.

Kelly, Louise, comp. *Wichita County Beginnings.* Austin: Eakin Press, 1982.

King, Carl B., and Howard W. Risher, Jr. *The Negro in the Petroleum Industry.* Philadelphia: University of Pennsylvania Press, 1969.

King, John O. *Joseph Stephen Cullinan: A Study of Leadership in the Texas Petroleum Industry, 1897–1937.* Nashville: Vanderbilt University Press, 1970.

Kinley, Jesse D. *Call Kinley: Adventures of an Oil Well Fire Fighter.* Chickasha, Okla.: n.p., 1996.

Lambert, Paul, and Kenny A. Franks, eds. *Voices from the Oil Fields.* Norman: University of Oklahoma Press, 1984.

Larson, Henrietta M., and Kenneth Wiggins Porter. *History of Humble Oil and Refining Company: A Study in Industrial Growth.* New York: Harper and Brothers, 1959.

Leeston, Alfred M., John A. Chilton, and John C. Jacobs. *The Dynamic Natural Gas Industry.* Norman: University of Oklahoma Press, 1963.

Linsley, Judith Walker, and Ellen Walker Rienstra. *Beaumont A Chronicle of Promise.* Woodland Hills, Calif.: Windsor Publications, 1982.

Linsley, Judith Walker, Ellen Walker Rienstra, and Jo Ann Stiles. *Giant under the Hill.* Austin: Texas State Historical Association, 2002.

*Lloyds City Directories, Odessa, Texas, 1946–1960.*

Lynch, Gerald. *Roughnecks, Drillers, and Toolpushers: Thirty-Three Years in the Oil Fields.* Austin: University of Texas Press, 1987.

McBeth, Reid Sayers. *Pioneering the Gulf Coast: A Story of the Life and Accomplishments of Capt. Anthony F. Lucas.* n.p., n.d.

McCoy, Dorothy Abbott. *Oil, Mud, and Guts.* Brownsville, Tex.: privately published, 1977.

McDaniel, Robert W., with Henry C. Dethloff. *Pattillo Higgins and the Search for Texas Oil.* College Station: Texas A&M University Press, 1989.

Mills, Warner E., Jr. *Martial Law in East Texas.* Indianapolis: Bobbs-Merrill, 1960.

Moore, Richard R. *West Texas after the Discovery of Oil: A Modern Frontier.* Austin: Jenkins Publishing, 1971.

Myres, Samuel D. *The Permian Basin: Petroleum Empire of the Southwest.* Vol. 1, *Era of Discovery, from the Beginning to the Depression.* El Paso: Permian Press, 1973.

———. *The Permian Basin: Petroleum Empire of the Southwest.* Vol. 2, *Era of Advancement, from the Depression to the Present.* El Paso: Permian Press, 1977.

Olien, Roger M., and Diana Davids Olien. *Easy Money: Oil Promoters and Investors in the Jazz Age.* Chapel Hill: University of North Carolina Press, 1990.

———. *Life in the Oil Fields.* Austin: Texas Monthly Press, 1986.

———. *Oil Booms: Social Change in Five Texas Towns.* Lincoln: University of Nebraska Press, 1982.

———. *Oil and Ideology: The Cultural Construction of the American Petroleum Industry.* Chapel Hill: University of North Carolina Press, 2000.

———. *Wildcatters: Texas Independent Oilmen.* Austin: Texas Monthly Press, 1984.

Owen, Edgar Wesley. *Trek of the Oil Finders: A History of Exploration for Oil.* Tulsa: American Association of Petroleum Geologists, 1975.

Pope, Clarence. *An Oil Scout in the Permian Basin.* El Paso: Permian Press, 1972.

Rister, Carl Coke. *Oil: Titan of the Southwest.* Norman: University of Oklahoma Press, 1949.

Singerman, Philip. *An American Hero: The Red Adair Story.* Boston: Little, Brown, 1990.

Sinise, Jerry. *Black Gold and Red Lights.* Burnet, Tex.: Eakin Press, 1982.

Snider, L. C. *Oil and Gas in the Mid-Continent Fields.* Oklahoma City: Harlow Publishers, 1920.

Spellman, Paul N. *Spindletop Boom Days.* College Station: Texas A&M University Press, 2001.

Stanley F. *The Early Days of the Oil Industry in the Texas Panhandle.* Borger, Tex.: Hess, 1973.

———. *The Phillips, Texas Story.* Nazareth, Tex.: F. Stanley, 1975.

———. *The Tex Thornton Story.* Nazareth, Tex.: F. Stanley, 1975.

Sterling, William Warren. *Trails and Trials of a Texas Ranger.* Norman: University of Oklahoma Press, 1968.

Stowe, Estha Briscoe. *Oil Field Child.* Fort Worth: Texas Christian University Press, 1989.

Warner, C. A. *Texas Oil and Gas since 1543.* Houston: Copano Bay Press, 2007.

Weaver, Bobby D., ed. *Panhandle Petroleum.* Canyon, Tex.: Panhandle-Plains Historical Society, 1982.

Whisenhunt, Donald W., ed. *Texas: A Sesquicentennial Celebration.* Austin: Eakin Press, 1984.

White, John H. *Borger, Texas: A History and the Real Facts about the Most Talked about Town in Texas and the Southwest.* Waco: Texian Press, 1973.

Woodward, Don. *Black Diamonds! Black Gold! The Saga of Texas Pacific Coal and Oil Company.* Lubbock: Texas Tech University Press, 1998.

## PERIODICALS

*Chronicles of Oklahoma*
*Gulf Coast Oil News*
*The Lamp*
*National Petroleum News*
*Oil and Gas Journal*
*Oil Investors Journal*
*Oil Weekly*
*Panhandle-Plains Historical Review*
*Permian Historical Annual*
*Petroleum Age*
*Petroleum Engineer*
*Pioneer America*
*Saturday Evening Post*
*Southwestern Historical Quarterly*
*Texas Monthly*
*West Texas Historical Association Yearbook*
*World Oil*

## NEWSPAPERS

*Amarillo Daily News*
*Amarillo Daily Tribune*
*Amarillo Globe*
*Amarillo Sunday News*
*Amarillo Sunday News and Globe*
*Andrews County News*
*Beaumont Age*
*Beaumont Enterprise*
*Beaumont Journal*
*Borger Daily Herald*
*Borger News-Herald*
*The Canyon News* (Canyon, Texas)
*Daily Panhandle* (Amarillo)
*Daily Times* (Wichita Falls)
*Dallas Morning News*
*The Dublin Progress* (Dublin, Texas)
*The Evening Post* (Amarillo, Texas)
*Fort Worth Star Telegram*
*Henderson Daily News* (Henderson, Texas)
*Houston Post*
*Hutchinson County Herald*
*Kansas City Star*
*Kilgore News-Herald*
*Midland Reporter-Telegram*
*Odessa American*
*Odessa News-Times*
*Odessa Times*
*Pecos Enterprise and Gusher*
*Ranger Daily Times*
*Ranger Times*
*San Angelo Standard-Times*
*Snyder Daily News*
*The Snyder News*
*The Snyder Signal*
*The Southwestern Plainsman* (Amarillo, Texas)
*Texas Evangel*
*Weekly Times* (Wichita Falls)
*Winkler County News*
*Winkler County Times*

## REFERENCE WORKS

Handbook of Texas Online. At http://www.tshaonline.org/handbook/online/.

Tyler, Ron, ed. *The New Handbook of Texas*. 6 vols. Austin: Texas State Historical Association, 1996.

## UNPUBLISHED WORKS

Harvey, Nyla. "History of Borger, Texas." Master's thesis, West Texas State College, 1953.

Horton, Finus Wade. "A History of Ector County, Texas." Master's thesis, University of Texas, 1950.

Huddleston, John D. "Good Roads for Texas: A History of the Texas Highway Department, 1917 to 1947." PhD diss., Texas A&M University, 1981.

Knoy, Robbie. "The Oil Boom in Snyder, Texas." Master's thesis, Hardin Simmons University, 1950.

Kreidler, Tai. "The Offshore Oil Industry." PhD diss., Texas Tech University, 1997.

Marshall-Gray, Paula. "Paradise Lost? Paradise Found? A Re-Collective History of Oilfield Culture in a West Texas Company Town." PhD diss., Texas Tech University, 2008.

Palmer, Derwin. "A History of the Desdemona Oil Boom." Master's thesis, Hardin Simmons University, 1938.

Parker, Albert Raymond. "Life and Labor in the Mid-Continent Oil Fields, 1859–1945." PhD diss., University of Oklahoma, 1951.

Phillips, Betty J. "Grand Old Man of the Panhandle Oil Industry; Jacob Rice Phillips." Master's thesis, West Texas State University, 1970.

Smith, Julia Cauble. "The Early Development of the Big Lake Oil Field, Reagan County, Texas." Master's thesis, University of Texas of the Permian Basin, 1986.

## INTERVIEWS

Abell, George, interviewed by Samuel D. Myres, September 23, 1971. PBPM.

Adams, Albert, interviewed by Bobby H. Johnson, August 12, 1970. SFA.

Adams, Mrs. Albert, interviewed by Bobby H. Johnson, August 13, 1970. SFA.

Akin, Jack, interviewed by Bobby Weaver, November 8, 1978. SWC.

Akins, Cullen, interviewed by Samuel D. Myres, June 17, 1970. PBPM.

Alexander, John D., interviewed by Richard Mason, October 10, 1982. SWC.

Allman, Mr. and Mrs. William W., interviewed by Paul Patterson, May 2, 1968, SWC.

Allman, William W., interviewed by Samuel D. Myres, June 19, 1970. PBPM.

Anderson, J. H., interviewed by Mrs. Maude Rose, January 29, 1960. CAH.

Angstadt, Carl, interviewed by Mody Boatright, August 4, 1952. CAH.

Armstrong, Rev. Pat M., interviewed by Fred Carpenter, September 2, 1970, SWC.
Armstrong, William L., interviewed by William A. Owens, August 24, 1953. CAH.
Ash, William Franklin "Hot Shot," interviewed by Roger Olien, March 2, 1979. UTPB.
Ashburn, Bruce, interviewed by Paul Patterson, April 19, 1968. SWC.
Autry, Eddie, interviewed by Richard Mason, January 27, 1980. SWC.
Avary, J. C., interviewed by Richard Mason, April 1, 1981. SWC.
Baines, Gordon, interviewed by Richard Mason, August 21, 1980, SWC.
Baird, Cal, remembrances, n.d. PPHM.
Barfoot, Lillian Hull, interviewed by Lou Scoggins, May 13, 1967. SWC.
Baker, Don, interviewed by Mary E. Davidson, April 27, 1976. PPHM.
Baldwin, Capt. B. C., interviewed by Bobby H. Johnson, September 19, 1970. SFA.
Barfield, Plummer M., interviewed by William A. Owens, August 1, 1952. CAH.
Bartlett, N. D., remembrances, July 31, 1936. PPHM.
Barton, Clyde, interviewed by Samuel D. Myres, August 15, 1970. PBPM.
Beane, E. N., interviewed by Samuel D. Myres, June 19, 1970. PBPM.
Beaver, E. A., interviewed by Richie Cravens, July 22, 1976. SWC.
Bellinghausen, Henry, interviewed by Bobby Weaver, August 29, 1980. PPHM.
Bentley, George R., interviewed by Samuel D. Myres, May 20, 1970. PBPM.
Bickley, Cecil, interviewed by Bobby Weaver, November 8, 1978. SWC.
Bickley, Cecil, interviewed by Richard Mason, August 24, 1982. SWC.
Bloodworth, Bert C., interviewed by Bobby Weaver, January 17, 1978. SWC.
Bobo, Dr. Tom C., interviewed by Samuel D. Myres, 1970, PBPM.
Boren, Walter, interviewed by Richard Mason, January 23, 1980. SWC.
Bosworth, Mr. and Mrs. Guy, interviewed by Paul Patterson, June 12, 1968. SWC.
Bott, Ted, interviewed by Bobby Weaver, November 2, 1978. SWC.
Bowman, John, interviewed by Bobby Weaver, 1958. Author's collection.
Boyd, Hiley, Jr., interviewed by David B. Gracey, II, May 20, 1969. SWC.
Boydston, Clyde, interviewed by Joanna Shurbet, March 9, 1976. SWC.
Brackens, E. E., interviewed by Samuel D. Myres, August 16, 1970. PBPM.
Brightman, T. C. "Tex," interviewed by Paul Patterson, May 16, 1968. SWC.
Brooks, Lewis W., interviewed by Fred A. Carpenter, 1969. SWC.
Brown, J. D., interviewed by Bobby Weaver, 1960. Author's collection.
Brown, M. K., interviewed by F. H. Doucette, March 21, 1944. PPHM.
Bruce, Lloyd, interviewed by Richie Cravens, November 26, 1976. SWC.
Bruce, Lloyd E., interviewed by Richard Mason, January 15, 1981. SWC.
Bryan, Mr. and Mrs. Burt, interviewed by S. D. Thompson, April 7, 1975. PPHM.
Bryant, W. H. "Bill," interviewed by William A. Owens, July 29, 1952. CAH.
Bryant, Mrs. Paul, interviewed by Bobby Weaver, April 12, 1978. SWC.
Burrell, Ruth Clay, interviewed by Judy McKowan, November 1, 1975. PPHM.
Burton, "Red," interviewed by Fred A. Carpenter, 1969. SWC.
Butz, Mr. and Mrs. Karl, interviewed by Fred A. Carpenter, September 23, 1975. SWC.

Campbell, Mrs. Seth (Neva), interviewed by Bobby Weaver, March 13, 1978. SWC.
Caraway, Mrs. John, interviewed by Fred Carpenter, June 18, 1970. SWC.
Carey, Gus, interviewed by Fred Carpenter, August 9, 1972. SWC.
Carsey, J. Ben, interviewed by Samuel D. Myres, October 9, 1970. PBPM.
Carter, Clyde, interviewed by Bobby Weaver, 1958. Author's collection.
Case, C. W., interviewed by Mary E. Davidson, November 10, 1975. PPHM.
Casey, Charles A., interviewed by Bobby H. Johnson, August 7, 1970. SFA.
Castleberry, Frank, interviewed by Mary E. Davidson, February 12, 1976. PPHM.
Castleberry, Frank, interviewed by Darryl Hemphill, December 1, 1981. PPHM.
Chalmers, Burns, interviewed by Charles L. Wood, July 3, 1980. SWC.
Champion, Frank, interviewed by Ruth Hosey, n.d. SWC.
Chancellor, Charles W., interviewed by Samuel D. Myres, October 29, 1971. PBPM.
Chase, Sharon, interviewed by Mary E. Davidson, May 10, 1976. PPHM.
Chiles, H. E. "Eddie," interviewed by Samuel D. Myres, February 1971. PBPM.
Clayton, George, interviewed by Samuel D. Myres, July 17, 1970. PBPM.
Clewell, Evelyn, interviewed by Richard Mason, April 15, 1982. SWC.
Cline, Walter, interviewed by Mody Boatright and others, August 18, 1952. CAH.
Collins, Mrs. Ida S., interviewed by Jean A. Paul, August 26, 1958. SWC.
Collyns, W. H. "Bill," interviewed by Richard Mason, January 25, 1980. SWC.
Conn, Mrs. Daisy, interviewed by Ann Clark, August 7, 1967, SWC.
Conselman, Frank B., interviewed by Richard Mason, April 12, 1982. SWC.
Cooper, S. O., interviewed by Samuel D. Myres, May 13, 1970. PBPM.
Cooper, Mr. and Mrs. Spurgeon, interviewed by Fred A. Carpenter, November 16, 1970. SWC.
Cotton, H. C. "Doc," interviewed by Bobby Weaver, November 7, 1978. SWC.
Cotton, William Edward, interviewed by William A. Owens, May 23, 1956. CAH.
Cowden, Elliott, interviewed by Samuel D. Myres, September 16, 1970. PBPM.
Cox, Mrs. F. L., interviewed by Mary E. Davidson, May 20, 1976. PPHM.
Cox, Wallace, interviewed by Richard Mason, January 1, 1982. SWC.
Coyle, Benjamin "Bud," interviewed by William A. Owens, July 29, 1952. CAH.
Craig, E. F., interviewed by Paul Patterson, n.d. SWC.
Crawford, Robert H., interviewed by Bobby Weaver, January 1, 1978. SWC.
Cullen, Robert A. "Bob," interviewed by Bobby Weaver, July 30, 2009. Author's collection.
Cullum, Landon Hayes, interviewed by Robert O. Stephens, January 3, 1959. CAH.
Cunningham, Mrs. Jimmie, interviewed by Mary E. Davidson, February 19, 1976. PPHM.
Daniels, Mr. and Mrs. V. B., interviewed by William A. Owens, July 3, 1952. CAH.
Davis, Charlie, interviewed by Richard Mason, February 12, 1981. SWC.
Davis, Dr. D. W., interviewed by William A. Owens, July 18, 1953. CAH.
Davis, Mr. and Mrs. Homer J. "Jack," interviewed by Bobby H. Johnson, August 21, 1970. SFA.

Deane, Early C., interviewed by William A. Owens, July 10, 1953. CAH.
Deer, Claud, interviewed by William A. Owens, July 7, 1952. CAH.
Dillard, A. R., interviewed by Mody Boatright, September 5, 1953. CAH.
Donnelly, George A., interviewed by Samuel D. Myres, February 1971. PBPM.
Donohoe, James, interviewed by William A. Owens, August 1, 1952. CAH.
Dixon, Crown N., interviewed by Bobby H. Johnson, August 21, 1970. SFA.
Drake, A. E., interviewed by Richie Cravens, December 29, 1975. SWC.
Duke, Jerry, interviewed by Bobby Weaver, October 14, 1992. Author's collection.
Dunn, Frank, interviewed by William A. Owens, March 23, 1955. CAH.
Edmonds, Cline, interviewed by Mary E. Davidson, April 15, 1976. PPHM.
Edwards, John, interviewed by Bobby Weaver, June 26, 1982. Author's collection.
Elder, Mrs. Eugene C. Kennedy, interviewed by Bobby H. Johnson, 1970. SFA.
Estes, Carl D., interviewed by Samuel D. Myres, May 20, 1970. PBPM.
Everett, Perry "Peb," interviewed by Bobby Weaver, December 6, 1978. SWC.
Fair, Charles "Chick," interviewed by Fred A. Carpenter, February 16, 1972. SWC.
Field, Mr. and Mrs. N. L., interviewed by Bobby H. Johnson, August 20, 1970. SFA.
Finger, George, interviewed by Mary E. Davidson, November 19, 1975. PPHM.
Flaughter, Mrs. Aline, interviewed by Mary E. Davidson, April 13, 1976. PPHM.
Foshee, Gladys B., interviewed by Bobby H. Johnson, August 30, 1970. SFA.
Foster, Tom E., interviewed by Bobby H. Johnson, August 4, 1970. SFA.
Fowler, Rayford, interviewed by Richard Mason, February 2, 1982. SWC.
Friend, E. M., interviewed by Mody Boatright, September 4, 1953. CAH.
Gebing, Mrs. Grace, interviewed by Richard Mason, February 23, 1982. SWC.
Gibson, Everett and Lillie, interviewed by Richard Mason, August 21, 1980. SWC.
Glass, Glenn, interviewed by Bobby Weaver, 1958. Author's collection.
Golding, Joe, interviewed by Neil Sapper, August 21, 1975. SWC.
Graybeal, Joseph W., interviewed by Samuel D. Myres, n.d. PBPM.
Greenhaw, Walter, interviewed by Richard Mason, October 30, 1981. SWC.
Griffen, F. E., interviewed by Jeff Townsend, August 22, 1972. SWC.
Hagy, Lawrence, interviewed by Neil Sapper, August 1, 1975. SWC.
Hagy, Lawrence, interviewed by David Nail, January 14, 1976. SWC.
Hagy, Lawrence, interviewed by James H. Gamberton, August 18, 1978. PPHM.
Haigh, Berte E., interviewed by Samuel D. Myres, August 21, 1969. PBPM.
Halsey, Harold, interviewed by Maude Ross, January 27, 1960. CAH.
Hamill, Allen W., interviewed by William A. Owens, September 2, 1952. CAH.
Hamill, Curt G., interviewed by William A. Owens, August 17, 1952. CAH.
Hamilton, Frank, interviewed by Mody Boatright, September 29, 1952. CAH.
Hamrick, Mr. and Mrs. M. R., interviewed by Charles Townsend, April 7, 1967. SWC.
Harder, Ware, interviewed by Mary E. Davidson, April 8, 1976. PPHM.
Harding, "Whitey," interviewed by Bobby Weaver, 1958. Author's collection.
Harris, T. B. "Buck," interviewed by Bobby Weaver, June 15, 1978. SWC.

Hayes, W. C., interviewed by Samuel D. Myres, n.d. PBPM.
Hedberg, H. A., interviewed by Samuel D. Myres, June 26, 1970, PBPM.
Heeter, Inez, interviewed by Richard Mason, October 30, 1981. SWC.
Henderson, Charles, interviewed by Fred A. Carpenter, October 19, 1973. SWC.
Henderson, Dick, interviewed by Richie Cravens, June 4, 1976. SWC.
Henshaw, Olive Elizabeth, interviewed by Cindy Hills Melanson, December 8, 1982. PPHM.
Hickox, Horace W., interviewed by Don Abbe, March 5, 1982. PPHM.
Higginbotham, Mack, interviewed by Fred A. Carpenter, 1969. SWC.
Hobson, R. R., interviewed by William A. Owens, May 15, 1956. CAH.
Hogan, Mr. and Mrs. Thomas C., interviewed by Paul Patterson, May 1, 1968. SWC.
Hollingsworth, Iris, interviewed by Clifford Ashby, September 5, 1982. SWC.
Holt, Jerry, interviewed by Bobby Weaver, June 10, 1980. Author's collection.
Hopper, Mr. and Mrs. A. C., interviewed by Bobby H. Johnson, August 8, 1970. SFA.
Hudson, W. M., interviewed by W. M. Hudson Jr., September 18,1952. CAH.
Huffman, Berl, interviewed by Perry McWilliams, August 15, 1971. SWC.
Huffman, Berl, interviewed by Bobby Weaver, May 21, 1979. SWC.
Hughes, C. Don, interviewed by Amelia S. Nelson, November 4, 1979. PPHM.
Hughes, C. Don, interviewed by Bobby Weaver, August 23, 1982. PPHM.
Hughes, T. M., interviewed by Paul Patterson May 1, 1968. SWC.
Hull, Bert E., interviewed by William A. Owens, August 24, 1953. CAH.
Hunt, O. B., interviewed by Dennis R. Brown, November 9, 1980. PPHM.
Hunt, W. C. "Bill," interviewed by Bobby Weaver, December 12, 1977. SWC.
Isett, Phillip, interviewed by Jeff Townsend, January 25, 1974. SWC.
Ivey, Buff and LaCosta, interviewed by Bobby Weaver, November 7, 1978. SWC.
Jennings, Fred, interviewed by William A. Owens, June 19, 1952. CAH.
Johnson, Claude E. "Peeper," interviewed by Bobby Weaver, July 29, 2009. Author's collection.
Johnson, Harold R., interviewed by Bobby H. Johnson, August 15, 1970. SFA.
Johnson, Herbert Lee "Peewee," interviewed by Bobby Weaver, 1959. Author's collection.
Johnson, M. A. "Curley," interviewed by Mody Boatright, August 27, 1952. CAH.
Johnson, M. T., interviewed by Marjory L. Morris, June 27, 1972. PPHM.
Johnson, Mr. and Mrs. R. L., interviewed by Mary E. Davidson, April 27, 1976. PPHM.
Johnson, W. R., interviewed by J. M. Skaggs, April 1, 1968. SWC.
Jones, Otto F., interviewed by Linda Webb, August 6, 1974. SWC.
Jowell, Holt, interviewed by Samuel D. Myres, September 16, 1970. PBPM.
Justin, Miss Eunice, interviewed by Charles Townsend, June 24, 1969. SWC.
Keith, Clarence "Sonny," interviewed by Bobby Weaver, 1957. Author's collection.
Keith, Gus, interviewed by Bobby Weaver, March 3, 1982. PPHM.
Keith, Jewel, interviewed by Mary E. Davidson, April 27, 1976. PPHM.

Keith, Jewel, interviewed by Mary E. Davidson, May 6, 1976. PPHM.
Kelly, Harry, interviewed by Bobby Weaver, December 7, 1978. SWC.
Kelton, R. W. "Buck," interviewed by Paul Patterson, August 1, 1972. SWC.
Kennedy, R. S., interviewed by Mody Boatright, July 30, 1952. CAH.
Killiam, O. W., interviewed by William A. Owens, May 7, 1956. CAH.
Kinnear, James William, interviewed by William A. Owens, July 24, 1953. CAH.
Kinnear, James William, interviewed by William A. Owens, August 16, 1953. CAH.
Klink, Mr. and Mrs. Anton, interviewed by Bobby Weaver, January 17, 1978. SWC.
Klinke, Esther, interviewed by Bobby Weaver, August 29, 1980. PPHM.
Knight, Jack, interviewed by Mody Boatright, September, 1952. CAH.
Knight, Jack, interviewed by Mary E. Davidson, April 18, 1976. PPHM.
Kostiha, Walter, interviewed by Richard Mason, March 5, 1980. SWC.
Kunkel, Max, interviewed by Bobby Weaver, January 18, 1978. SWC.
Lacy, Joe D., interviewed by Bobby Weaver, August 15, 1980. PPHM.
Landers, Bradford, interviewed by Bobby Weaver, March 23, 1979. SWC.
Langton, Mrs. F. E., interviewed by Charles Townsend, April 7, 1967. SWC.
Lantron, E. L., interviewed by Mody Boatright, September 8, 1952. CAH.
Lapin, Olga Herrman, interviewed by Bobby H. Johnson, August 31, 1970. SFA.
Lasater, E. C., interviewed by R. M. Hays, n.d. CAH.
Laughlin, Charles, interviewed by Richard Mason, September 28, 1981. SWC.
Laughlin, Lena, interviewed by Richard Mason, May 20, 1981. SWC.
Lawrence, I. D. "Cotton," interviewed by Bobby Weaver, 1958. Author's collection.
Lawson, O. G., interviewed by Mody Boatright, July 28, 1952. CAH.
Leach, Walter, interviewed by Fred A. Carpenter, September 19, 1974. SWC.
Leslie, Michael, interviewed by Bobby Weaver, September 4, 2008. Author's collection.
Littlejohn, Margaret, interviewer unknown, April 15, 1957. SWC.
Lively, Dan, interviewed by Paul Patterson, May 23, 1968. SWC.
Locklin, Dee, interviewed by Samuel D. Myres, February 13, 1970. PBPM.
Lomax, Thornton, interviewed by Richard Mason, January 15, 1982. SWC.
Longcope, Charles, interviewed by Fred A. Carpenter, 1969. SWC.
Love, Mary Florey, interviewed by Bobby H. Johnson, July 30, 1970. SFA.
Maddox, Bing, interviewed by Don Abbe, February 9, 1982. PPHM.
Martin, Fred, interviewed by Richard Mason, September 10, 1981. SWC.
Matteson, Ed. R., interviewed by Mody Boatright, June 19, 1953. CAH.
McCarthy, Glenn, interviewed by William A. Owens, July 29, 1953. CAH.
McClelland, C. C., interviewed by Mody Boatright, September 9, 1952. CAH.
McDonald, Archie Jones, interviewed by Bobby Weaver, October 8, 1977. SWC.
McGinnis, William, interviewed by Richard Mason, August 14, 1980. SWC.
McKinney, C. L., interviewed by Samuel D. Myres, September 23, 1971. PBPM.
McLaughlin, Ralph B., interviewed by William A. Owens, May 28, 1956. CAH.
McPhee, John, interviewed by Fred A. Carpenter, April 1, 1976. SWC.

McWhorter, Ralph, interviewed by Bobby Weaver, March 14, 1978. SWC.
Mendenhall, W. H. "Toby," interviewed by Bobby Weaver, March 13, 1985 (PPHM).
Merchant, Lawrence, interviewed by Richard Mason, February 10, 1982. SWC.
Michaels, Herman "Little Red," interviewed by Bobby Weaver, 1958. Author's collection.
Middleton, Wayne "Lefty," interviewed by Steven Gamble, July 12, 1976. SWC.
Miller, Mrs. Doris, interviewed by Fred A. Carpenter, October 6, 1975. SWC.
Mills, Buster, interviewed by Richie Cravens, June 4, 1976. SWC.
Mirus, Carl F., interviewed by William A. Owens, April 4, 1956. CAH.
Montgomery, Charlotte Baker, interviewed by Bobby H. Johnson, July 16, 1985. SFA.
Morley, Harold T., interviewed by Samuel D. Myres, October 20, 1970. PBPM.
Morris, C. Lee, interviewed by Richard Mason, April 22, 1981. SWC.
Morton, "Junior," interviewed by Bobby Weaver, 1958. Author's collection.
Moser, Mrs. Mary, interviewed by Mary E. Davidson, May 20, 1976. PPHM.
Myers, Orville, interviewed by Samuel D. Myres, October 29, 1971. PBPM.
Neeley, Tom, interviewed by Richard Mason, April 8, 1982. SWC.
Nichols, H. P., interviewed by Mody Boatright, October 11, 1952. CAH.
Nichols, Dr. James L., interviewed by Bobby H. Johnson, August 11, 1970. SFA.
Nolan, Mrs. Jack, interviewed by Bobby Weaver, January 13, 1978. SWC.
O'Donohoe, John F., interviewed by Mody Boatright, September 4, 1953. CAH.
Page, Mr. and Mrs. D. W., interviewed by Mary E. Davidson, April 22, 1976. PPHM.
Paige, Sidney, interviewed by William A. Owens, June 7, 1954. CAH.
Paramore, Harry R., interviewed by William A. Owens, July 2, 1953. CAH.
Parker, Jackson E., interviewed by Fred A. Carpenter, March 6, 1972. SWC.
Parker, S. A., interviewed by Bobby H. Johnson, August 13, 1970. SFA.
Parrack, Earl, interviewed by Bobby Weaver, February 15, 1979. SWC.
Paschall, Burk, interviewed by Mody Boatright, July 30, 1952. CAH.
Patterson, Alexander Balfour, interviewed by William A. Owens, August 4, 1953. CAH.
Patterson, Paul, interviewer unknown, September 12, 1973. SWC.
Peacock, Floyd, interviewed by Jeff Townsend, August 22, 1972. SWC.
Peebles, Jim S., interviewed by Samuel D. Myres, July 16, 1970. PBPM.
Perkins, Joseph J., interviewed by Mody Boatright, September 4, 1953. CAH.
Petsch, Colonel Alfred, interviewed by Fred A. Carpenter, September 3, 1975. SWC.
Phillips, Loyce, interviewed by Bobby H. Johnson, July 29, 1970. SFA.
Philp, William Joseph, interviewed by William A. Owens, July 17, 1953. CAH.
Pickens, T.B., Sr., interviewed by Mary E. Davidson, June 19, 1978. PPHM.
Pickle, Joe, interviewed by Richard Mason, February 27, 1980. SWC.
Pickrell, Frank, interviewed by Samuel D. Myres, August 21, 1969. PBPM.
Pitner, K. O., interviewed by Jeff Townsend, August 22, 1972. SWC.
Pratt, Wallace, interviewed by E. DeGolyer, March 2, 1945 (DeGolyer Papers, DeGolyer Library, Southern Methodist University, Dallas, Texas).

Rainosek, E. P. "Jack," interviewed by Samuel D. Myres, June 17, 1970. PBPM.
Ramer, George W., interviewed by Samuel D. Myres, February 12, 1970. PBPM.
Randel, C. M., interviewed by Fred A. Carpenter, n.d. SWC.
Randolph, Charles W., interviewed by William A. Owens, spring 1955. CAH.
Randolph, Frank E., interviewed by Fred Carpenter, December 11, 1975. SWC.
Rathke, H. E., interviewed by Mody Boatright, September 13, 1953. CAH.
Redman, Frank, interviewed by William A. Owns, July 20, 1953. CAH.
Rees, Mrs. Wade, interviewed by Fred A. Carpenter, March 13, 1972. SWC.
Richardson, Leo, interviewed by Elmer Kelton, March 29, 1976. SWC.
Riley, Mr. and Mrs. Leonard, interviewed by Mary E. Davidson, April 22, 1976. PPHM.
Roberson, Dock, interviewed by Richard Mason, February 6, 1980. SWC.
Roberts, Hardeman, interviewed by William A. Owens, April 26, 1956. CAH.
Roberts, James, interviewed by Jeff Townsend, August 22, 1972. SWC.
Robinson, James and Leota, interviewed by Richard Mason, May 17, 1982. SWC.
Robinson, Raymond A., interviewed by Bobby H. Johnson, August 20, 1971. SFA.
Robison, Polk, interviewed by Bobby Weaver, May 16, 1979. SWC.
Ross, James F., interviewed by Mody Boatright, September 4, 1956. CAH.
Ross, Mrs. Sam W., interviewed by Bobby H. Johnson, 1970, SFA.
Ruby, Mrs. Claude, interviewed by Mary E. Davidson, November 7, 1975. PPHM.
Rumble, Jo Ann, reminiscence, n.d. PPHM.
Rush, J. A., interviewed by Mody Boatright, September 11, 1956. CAH.
Russell, Frank, interviewed by Bobby Weaver, January 17, 1979. SWC.
Rust, John, interviewed by Mody Boatright, September 12, 1952. CAH.
Sandefer, J. D., interviewed by David Murrah, September 8, 1972. SWC.
Sanderson, Phillip A., interviewed by Bobby H. Johnson, August 19, 1970. SFA.
Sappington, Mrs. Tom, interviewed by Mary E. Davidson, May 6, 1976. PPHM.
Schlicher, Max Theodore, interviewed by William A. Owens, July 1, 1953. CAH.
Scott, Jack, interviewed by Richard Mason, March 12, 1981. SWC.
Sherman, Max, interviewed by Mary E. Davidson, July 20, 1978. PPHM.
Sherwood, Mrs. H. E., interviewed by Bobby H. Johnson, 1970. SFA.
Sill, P. O., interviewed by Samuel D. Myres, April 12, 1970. PBPM.
Sitton, F. L., interviewed by Richard Mason, September 29, 1981. SWC.
Skeeters, L. L., interviewed by Bobby H. Johnson, 1970. SFA.
Skeeters, Mrs. L. L. Tullos, interviewed by Bobby H. Johnson, 1970. SFA.
Sloop, Dr. H. C., interviewed by William A. Owens, March 25, 1955. CAH.
Smith, B. H., interviewed by Bobby H. Johnson, 1970. SFA.
Smith, Barney, interviewed by Bobby Weaver, 1974. Author's collection.
Southern, Bernadette, interviewed by Bobby Weaver, November 8, 1978. SWC.
Spear, Mrs. E. H., interviewed by Bobby H. Johnson, August 13, 1970. SFA.
Stamey, O. L., interviewed by Richie Cravens, November 26, 1976. SWC.
Starl, Orlin, Sr., interviewed by Bobby Weaver, March 16, 1978. SWC.

Stivers, Mr. and Mrs. Bert, interviewed by William A. Owens, August 26, 1953. CAH.
Stout, Arthur, interviewed by Samuel D. Myres, n.d. PBPM.
Studdard, George B., interviewed by Richard Mason, February 13, 1981. SWC.
Sudduth, Miss Nettie, interviewed by Fred Carpenter, June 18, 1970. SWC.
Summers, F. E. "Ellis," interviewed by Richard Mason, March 31, 1981. SWC.
Swanson, F. G., interviewed by R. M. Hayes, February 20, 1955. CAH.
Swanson, John D., interviewed by Fred Carpenter, December 17, 1974. SWC.
Taylor, Bill N., interviewed by Bobby H. Johnson, August 13, 1970. SFA.
Taylor, Grimm, interviewed by Bobby Weaver, October 8, 1877. SWC.
Ten Eyck, W. E. "Bill," interviewed by Fred A. Carpenter, April 1, 1976. SWC.
Thomas, A. J., interviewed by Mrs. Maude Ross, January 29, 1960. CAH.
Thompson, Eugene "Shorty," interviewed by Fred A. Carpenter, November 5, 1975. SWC.
Thompson, Fritz, interviewed by Mary E. Davidson, April 13, 1976. PPHM.
Thompson, Ralph, interviewed by Bobby Weaver, 1965. Author's collection.
Townes, Judge Edgar Eggleston, interviewed by William A. Owens, June 21, 1952. CAH.
Twichell, W. D. Family, interviewed by David Murrah, May 7, 1973. SWC.
Velaz, Nick, interviewed by Chris Johnson, March 21, 1976. PPHM.
Vincent, Clifford, interviewed by Jesse Baker, January, 1970. PPHM.
Waddington, Johnny, interviewed by Richard Mason, January 4, 1980. SWC.
Wagner, "Skeet," interviewed by Bobby Weaver, spring 1983. Author's collection.
Walker, Bill, interviewed by Bobby Weaver, 1963. Author's collection.
Walker, George F., interviewed by Fred Carpenter, September 24, 1974. SWC.
Ware, Matt, interviewed by Bobby Weaver, July 24, 2009. Author's collection.
Ware, Penn, interviewed by Bobby Weaver, 1955. Author's collection.
Warren, David, interviewed by Mary E. Davidson, April 15, 1976. PPHM.
Watson, W. Lee, interviewed by Richard Mason, January 20, 1981. SWC.
Weaver, Carl, interviewed by Paul Patterson, July 5, 1966. SWC.
Weaver, Robert S. "Bob," interviewed by Bobby Weaver, 1974. Author's collection.
Webb, J. R., interviewed by Mody Boatright, August 1, 1952. CAH.
Webb, Mr. and Mrs. Sam W., interviewed by William A. Owens, September 1952. CAH.
Weller, George Walker, interviewed by William A. Owens, August 22, 1953. CAH.
Whaley, Lance T., interviewed by Fred A. Carpenter, March 22, 1968. SWC.
Whatley, Albert, interviewed by Samuel D. Myres, July 18, 1970. PBPM.
Whatley, Richard, interviewed by Bobby Weaver, 1965. Author's collection.
Wheat, James J., interviewed by Samuel D. Myres, July 17, 1970. PBPM.
Wheeler, Bill, interviewed by Bobby Weaver, 1965. Author's collection.
Wilburn, Tony, interviewed by Richard Mason, April 1, 1981. SWC.
Williams, Clayton W., Sr., interviewed by Samuel D. Myres, June 4, 1969. PBPM.
Williams, James "Pete," Sr., interviewed by Samuel D. Myres, September 15, 1971. PBPM.
Williams, Paul, interviewed by Richie Cravens, July 7, 1977. SWC.
Williams, Tom J., interviewed by Fred A. Carpenter, 1969. SWC.

Williamson, C. L., interviewed by Jeff Townsend, September 16, 1972. SWC.
Windham, L. E. "Fuzz," interviewed by Samuel D. Myres, n.d. PBPM.
Winfrey, L. D., interviewed by Mrs. Robert A. Montgomery, August 2, 1959. CAH.
Witherspoon, Claude L., interviewed by Mody Boatright, June 27, 1953. CAH
Wolf, Mr. and Mrs. William, interviewed by Samuel D. Myres, February 12, 1970. PBPM.
Woods, H. D., interviewed by Richard Mason, April 2, 1982. SWC.
Wylie, Mr. and Mrs. P. D. "Pinky," interviewed by Richard Mason, March 5, 1980. SWC.
Wynn, John S., interviewed by William A. Owens, July 4, 1954. CAH.
Young, J. T. "Cotton." interviewed by Mody Boatright, August 15, 1952. CAH.
Zeigler, Jimmy, interviewed by Bobby Weaver, 1958. Author's collection.

# Index